Vignettes in Gravitation and Cosmology

T0324954

Vignettes in Gravitation and Cosmology

Editors

L. Sriramkumar
Harish-Chandra Research Institute, India

T. R. Seshadri
University of Delhi, India

World Scientific

NEW JERSEY · LONDON · SINGAPORE · BEIJING · SHANGHAI · HONG KONG · TAIPEI · CHENNAI

Published by

World Scientific Publishing Co. Pte. Ltd.
5 Toh Tuck Link, Singapore 596224
USA office: 27 Warren Street, Suite 401-402, Hackensack, NJ 07601
UK office: 57 Shelton Street, Covent Garden, London WC2H 9HE

British Library Cataloguing-in-Publication Data
A catalogue record for this book is available from the British Library.

ISBN-13 978-981-4322-06-5
ISBN-10 981-4322-06-7

Printed in Singapore by Mainland Press Pte Ltd.

Preface

This book is a unique collection of articles, written by former students and collaborators of T. Padmanabhan, on different aspects of gravitation and cosmology.

Padmanabhan, or Paddy, as he is affectionately known, constantly inspires us with his passion for physics, his intense enthusiasm for research, and his continued earnestness to work harder than any of his students. He has always been easily accessible to us and has played a significant role in shaping our outlook on physics and research. We hope that this collection of articles serves as a small token of our appreciation for the positive influence he has had on all of us. (At the end of the collection, we have listed the publications of all the contributors to this volume with Padmanabhan.)

Gravitation and cosmology are presently very active fields of research. The articles in this collection are focussed on topics that are of contemporary research interest, and they have been aimed at the level of graduate students. We believe that the articles will bridge the gaps that can exist between the discussions in the textbooks and the research level articles available on these topics.

We would like to thank all the contributors to this volume for their constant encouragement and support. We would also like to express our sincere gratitude to World Scientific, in particular, to Lakshmi Narayanan and her team, for guidance, support and infinite patience.

Editors: L. Sriramkumar and T. R. Seshadri

Contents

Chapter 1

Non-linear gravitational clustering in an expanding universe

Jasjeet Singh Bagla[1]

Harish-Chandra Research Institute,
Chhatnag Road, Jhunsi, Allahabad 211019, India.

Abstract: I outline some known facts in non-linear gravitational clustering in an expanding universe, with particular emphasis on aspects that are independent of any specific initial conditions. Several attempts have been made to understand these generic features of gravitational clustering. I will summarize some insights into these issues that have been gleaned from N-Body simulations and other models, that may help in solving this problem.

1.1 Introduction

Non-linear gravitational clustering in an expanding universe is of considerable interest in cosmology and galaxy formation. Galaxies form in highly overdense haloes where the gas can cool and fragment to form stars [1, 2, 3, 4]. The overdense haloes are believed to have formed through amplification of small fluctuations in the density distribution in the early universe [5, 6, 7, 8, 9].

Growth of density perturbations can be modeled analytically as long as the amplitude of perturbations is sufficiently small at the relevant scales. The amplitude of density perturbations is quantified by defining the density contrast: $\delta(\mathbf{r}, t) = \rho(\mathbf{r}, t)/\bar{\rho}(t) - 1$. Here $\rho(\mathbf{r}, t)$ is the density of matter at $\mathbf{r}$ at time t, and $\bar{\rho}(t)$ is the average density of the universe at time t. This is not possible if the amplitude of perturbations is large, as happens at small scales and at late times. We can use cosmological N-Body simulations to study the evolution of density perturbations in this regime. Other possibili-

---

[1]Current address: IISER Mohali, Transit campus: MGSIPAP complex, Sector 26, Chandigarh 160019, India. E-mail: jasjeet@iisermohali.ac.in.

ties include quasi-linear approximations and scaling relations. Quasi-linear approximations allow us to follow the evolution of density perturbations around the time when it becomes non-linear, whereas scaling relations give us more limited statistical information about the evolution of clustering.

We trace the large scale structure in the universe through observations of galaxies, and galaxies themselves represent very large overdensities. Therefore it is essential to follow the evolution of density perturbations deep into the non-linear regime. While simulations allow us to do this, these do not improve our understanding of the physical processes involved. This essentially means that we need to rerun simulations for each and every model. Approximation schemes and scaling relations do offer some insights but these must be calibrated with simulations.

In this article I review the insights that have been gained into the process of gravitational clustering with a combination of approximation schemes, scaling relations and cosmological N-Body simulations.

1.2   Gravitational clustering

Gravitational clustering is the evolution of a distribution of particles with only gravitational interaction between them. We assume that the system can be described in the Newtonian limit. This is a valid limit as long as the scales of interest are much smaller than the Hubble radius and the matter that clusters is non-relativistic [6]. The universe should not contain significant amount of radiation or relativistic species of particles if we are to use the Newtonian limit.

The growth of perturbations is then described by the coupled system of Euler's equation and Poisson equation in comoving coordinates along with mass conservation [5]:

$$\ddot{\mathbf{x}}_i + 2\frac{\dot{a}}{a}\dot{\mathbf{x}}_i = -\frac{1}{a^2}\nabla_i\varphi,$$

$$\nabla^2\varphi = 4\pi G a^2 \left(\rho - \bar{\rho}\right) = \frac{3}{2}\frac{\delta}{a^3}H_0^2\Omega_{nr},$$

$$\rho(\mathbf{x}) = \frac{1}{a^3}\sum_i m_i\delta_{D}(\mathbf{x} - \mathbf{x}_i). \tag{1.1}$$

It is assumed that the density field is generated by a distribution of particles: the ith particle has mass m_i and position $\mathbf{x}_i$. H_0 is the present value of Hubble constant, Ω_{nr} is the present density parameter for non-relativistic matter and a is the scale factor. The Hubble radius is c/H_0 and as mentioned above, the Newtonian limit is valid at scales much smaller than this. Here we consider the Einstein-de Sitter universe as the background, i.e., $\Omega_{nr} = 1$. This can be done without the loss of generality as far as non-linear gravitational clustering is concerned.

The equations given above can be reduced to a single non-linear differential equation for density contrast:

$$\ddot{\delta}_{\mathbf{k}} + 2\frac{\dot{a}}{a}\dot{\delta}_{\mathbf{k}} = \frac{3}{2}\frac{\delta_{\mathbf{k}}}{a^3}H_0^2 + A - B, \tag{1.2}$$

where

$$A = \frac{3}{4}\frac{1}{a^3}H_0^2 \sum_{\mathbf{k}' \neq 0, \mathbf{k}} \left[\frac{\mathbf{k}.\mathbf{k}'}{k'^2} + \frac{\mathbf{k}.(\mathbf{k} - \mathbf{k}')}{|\mathbf{k} - \mathbf{k}'|^2} \right] \delta_{\mathbf{k}'}\delta_{\mathbf{k}-\mathbf{k}'}$$

and

$$B = \frac{1}{M}\sum_i m_i \left(\mathbf{k}.\dot{\mathbf{x}}_i\right)^2 \exp\left[i\mathbf{k}.\mathbf{x}_i\right] \quad ; \quad M = \sum_i m_i.$$

The terms A and B are the non-linear coupling terms between different modes. B couples density contrasts in an indirect manner through velocities of particles $(\dot{\mathbf{x}}_i)$. The equation of motion still needs to be solved for a complete solution of this equation, or we can use some ansatz for velocities to make this an independent equation.

The problem of non-linear gravitational clustering in an expanding universe is essentially to solve the system of equations given above in situations where density contrast is much larger than unity. As physicists, just getting the solution is not satisfying, one must understand the solution as well.

1.2.1 *Linear approximation*

We begin by solving the system of equations in the regime where the amplitude of perturbations is small. This serves as a useful reference for further developments. In this regime, the equation for density contrast reduces to a second order linear and homogeneous ordinary differential equation:

$$\ddot{\delta}_{\mathbf{k}} + 2\frac{\dot{a}}{a}\dot{\delta}_{\mathbf{k}} - \frac{3}{2}\frac{\delta_{\mathbf{k}}}{a^3}H_0^2 = 0. \tag{1.3}$$

This has a growing solution and a decaying solution. The equation can be solved in a straightforward manner for the Einstein-de Sitter universe. In this case the scale factor $a(t) \propto t^{2/3}$ is the growing mode. This rate of growth is slow, certainly as compared to the rate of growth in absence of expansion. Almost equal time scales of expansion and gravitational collapse are responsible for this slow rate of growth of perturbations.

For a wide variety of models where we may have a universe with a mix of non-relativistic matter, curvature and cosmological constant, it can be shown that the Hubble parameter $H(t)$ is the decaying solution. Once we have one solution for such an equation, the Wronskian can then be used to compute the other solution [10]. The growing solution of Eqn. (1.3) is also referred to as the growing mode and is often denoted as $D_+(t)$.

The rate of growth is independent of scale. We have made a simplification by ignoring pressure that modifies the rate of growth of perturbations at very small scales. Modulo this approximation, it is clear that at early times the density perturbations evolve in a self-similar manner and that density contrast everywhere changes at the same rate. This clearly cannot be the correct solution at late times as δ is bounded from below. A more refined approach is required for understanding the evolution of density perturbations as we approach $|\delta| \simeq 1$. In principle one can use a higher order perturbation series in δ but clearly all of these fail around $|\delta| \simeq 1$ whereas we are interested in much stronger non-linearities.

1.2.2   *Quasi-linear approximations*

It is possible to rewrite the equation of motion in a form that helps motivate the form of quasi-linear approximations. We rewrite the equations that govern the evolution of perturbations using the growing mode $D_+(t)$ as the time variable.

$$\frac{d\mathbf{u}}{dD_+} = -\frac{3}{2}\frac{Q}{D_+}\left(\mathbf{u} - \mathbf{g}\right),$$

$$\nabla^2\psi = \left(\frac{\delta}{D_+}\right), \tag{1.4}$$

with

$$\mathbf{g} \equiv -\nabla\psi = -\frac{2}{3H_0^2\Omega_{nr}}\left(\frac{a}{D_+}\right)\nabla\varphi,$$

$$Q = \left(\frac{\bar{\varrho}}{\varrho_c}\right)\left(\frac{\dot{a}D_+}{a\dot{D}_+}\right)^2. \tag{1.5}$$

Here $\mathbf{u} = d\mathbf{x}/dD_+$ is the generalized velocity, ψ is the generalized gravitational potential and all other symbols have their usual meanings. We have written these equations for a general cosmological model to demonstrate that the treatment carries over to models other than the Einstein-de Sitter model.

It is apparent from the structure of the equations that in the linear regime ψ does not change with time. Further, it can be shown that in this regime du/dD_+ vanishes and $\mathbf{u} = \mathbf{g} = -\nabla\psi$. This can be interpreted in two ways, as we shall describe below.

The Zel'dovich approximation [11, 12] interprets the constancy of the velocity in the Lagrangian sense. It is assumed that the velocity $\mathbf{u}$ of a given particle does not change with time, and is set by the value of $\mathbf{g} = -\nabla\psi$ at its initial location. The location of the particle at a later time follows:

$$\mathbf{x}(t) = \mathbf{x}(t_{in}) - (D_+(t) - D_+(t_{in}))\nabla\psi(\mathbf{x}_{in}, t_{in}) \simeq \mathbf{q} - D_+(t)\nabla_{\mathbf{q}}\psi, \quad (1.6)$$

where $\mathbf{q}$ is the Lagrangian position $(\mathbf{x}(t_{in}))$ and we assume that the initial time is early enough that $D_+(t) \gg D_+(t_{in})$. The "inertial" motion of particles results from the near equality of the expansion and collapse time scales. Essentially the drag force due to expansion and the gravitational force of overdensities nearly balance each other. As long as shell crossing does not happen and the map described in this equation is one to one, we can use the requirement of mass conservation to estimate evolution of density.

$$1 + \delta = \frac{1}{|Det\{(1 - D_+\lambda_1)(1 - D_+\lambda_2)(1 - D_+\lambda_3)\}|}, \quad (1.7)$$

where λ_i are eigenvalues of the tidal tensor $\partial^2\psi/\partial q_i\partial q_j$. For small displacements, this expression is exact and we can recover linear theory. The expression becomes invalid at the time of shell crossing after which the Zel'dovich mapping is many to one. In the limit of small density perturbations, this approach should be regarded as the linear order Lagrangian perturbation theory. Extrapolation to situations where density contrast may be large is known as the Zel'dovich approximation. Comparisons with N-Body simulations show that the Zel'dovich approximation is remarkably successful and follows the actual trajectories remarkably well up to shell crossing. Further, it has been shown that if the power spectrum is truncated at wave numbers corresponding to non-linear scales then the Zel'dovich approximation reproduces the correct density and velocity distribution at large scales [13]. This aspect has been exploited by many to delink formation of haloes from the large scale motions and hence simulate evolution of large scale structure in a simple manner [14]. The simple form of Zel'dovich approximation also allows evaluation of the non-linear power spectrum [15, 16].

We can consider the eigenvalues to be ordered such that $\lambda_1 \geq \lambda_2 \geq \lambda_3$. It can be shown that for generic initial perturbations, equality occurs over a set of zero measure [12, 17]. Therefore at an arbitrary point, density diverges when $\lambda_1(\mathbf{q}) = 1/D_+$. That is, collapse leads to collapse onto a surface. Such surfaces of high density were termed as pancakes by Zel'dovich.

Beyond shell crossing particles cross the potential well and we expect these to turn back and remain confined around the overdense regions. Clearly this does not happen in the Zel'dovich approximation. This can be corrected at some level by going to a higher order Lagrangian perturbation theory but the issue of stability of pancakes in simulations remains unanswered in this approach.

The role of an expanding background in this context was highlighted by another quasi-linear approximation scheme [18, 19]. In this scheme, it is assumed that the gravitational potential evolves at the linear rate, i.e., $\psi(\mathbf{x})$ is assumed to remain constant. Even though any deepening of potential wells to confine particles is not taken into account, it was shown that this approximation leads to stabilization of pancakes. Clearly, this happens because of expansion of the universe as in absence of expansion particles will have very large amplitude of oscillations about the pancakes. Expansion provides a drag force that damps the infall velocities and this in turn lowers the oscillation amplitude of particles about pancakes.

1.3   In search of universalities

Gravitational clustering is essentially a process that amplifies density perturbations. If we describe the initial and the final density distributions in statistical terms, then one may ask whether the process of clustering can be described as a map. Such a map would be of utmost interest if it has some universal features that do not depend on the details of the initial density distribution as this may hold clues about the process of gravitational clustering.

Results of N-Body simulations of a variety of initial conditions [20] were used to illustrate one such universal aspect [21] of gravitational clustering in an expanding universe. It was shown that the final and the initial two point correlation are related through a non-local map, i.e., the initial correlation at a given scale leads to the final correlation at a smaller scale. Relation between the two scales was derived using the pair conservation equation [5, 21]. It is possible to relate the universal map for the correlation function if the average pair velocities depend only on the amplitude of the correlation function and not on initial conditions [22]. Why such a universality exists is not very clear though it can be motivated from ideas related to collapse of individual haloes [23, 24]. More detailed studies have shown that the relationship, while being close to universal, does retain some memory of initial conditions [25].

Even an approximate universality in the mapping from initial conditions to the final state is very useful. This allows us to estimate the evolution of correlation functions in a simple manner without resorting to numerical simulations. Indeed, it is possible to use the mapping to make the following statements [26]:

- The mapping reduces the differences in the slope of the correlation function at scales where its amplitude is of order unity, or larger. Hence we may say that gravitational clustering erases differences between initial conditions.
- The existence of a universal map does, at the same time, indicate that there is a one to one connection between the initial and the final state. Therefore the memory of initial conditions is not lost completely.

These are fairly powerful conclusions and it is clear that a better understanding of the mapping between initial and final conditions is required. Some progress in this direction has been made through the renormalized perturbation theory [27, 28] but this approach does not lead to a significant improvement in our understanding though it does provide a calculational tool that can be used to make predictions for the non-linear correlation function for a given initial correlation function.

We have stated above that the map between the initial and the final correlation function relates these at two different scales, the final scale being much smaller than the initial scale. This can be understood at an intuitive level as growth of perturbations is dominated by collapse around haloes, or through squeezing of matter in pancakes and filaments around underdense regions. As time goes on, more and more matter is confined to highly overdense haloes that take up a very tiny fraction of the total volume but contain most of the mass. In such a scenario, the growth of correlation function is dominated by infall [23]. As models based on infall predict a universal mapping between initial and final conditions, departures from universality must stem from the influence of already collapsed haloes on further infall. Therefore it is very important to understand the interaction of density perturbations at very different scales and specifically the influence of perturbations at small scales on larger scales. Several simulation studies have shown that this influence is weak [29, 30, 26, 31, 32], as expected from the existence of a nearly universal mapping between the initial and final correlation function. For further insight, we now turn to analytical models.

1.3.1 *Mode coupling: Effect of small scale perturbations*

It was shown by Peebles [33] that motions of matter within virialized dark matter haloes affect perturbations at larger scales in a very weak manner. Motions in systems that are not virialized lead to an affect of order $\mathcal{O}(k^4)$ at very small wave numbers. This has been confirmed in N-Body simulations [26]. A similar result is obtained if we rearrange matter while satisfying mass and momentum conservation, thus the form of the influence may not depend strongly on the fact that we are dealing with gravitational interaction.

Here we discuss a potentially interesting generalization of the results described above [34]. The mode coupling equation (Eq. 1.2) can be rewritten in the form [5]

$$\ddot{\delta}_{\mathbf{k}} + 2\frac{\dot{a}}{a}\dot{\delta}_{\mathbf{k}} = C_{\mathbf{k}} - D_{\mathbf{k}}, \tag{1.8}$$

$$C_{\mathbf{k}} = \frac{1}{M} \sum_j m_j \left[i\mathbf{k}. \left(-\frac{\nabla\phi_j}{a^2} \right) \exp\left(i\mathbf{k}.\mathbf{x}_j\right) \right], \tag{1.9}$$

$$D_{\mathbf{k}} = \frac{1}{M} \sum_j m_j \left(\mathbf{k}.\dot{\mathbf{x}}_j\right)^2 \exp\left[i\mathbf{k}.\mathbf{x}_j\right]. \tag{1.10}$$

The usual linear term in $\delta_{\mathbf{k}}$ is absorbed in $C_{\mathbf{k}}$. It has been shown that a single halo of particles in virial equilibrium does not contribute to $C_{\mathbf{k}} - D_{\mathbf{k}}$ at wave numbers much smaller than the inverse of the size of the halo. We consider the effect of interaction between two haloes in order to estimate the effect of mode coupling and transfer of power from small scales to large scales.

The first mode coupling term can be written as:

$$C_{\mathbf{k}} = \frac{1}{M} \sum_j m_j \left[i\mathbf{k}. \left(-\frac{\nabla\phi_j}{a^2} \right) \exp\left(i\mathbf{k}.\mathbf{x}_j\right) \right] = \frac{i}{M} \sum_j (\mathbf{k}.\mathbf{f}_j) \exp\left(i\mathbf{k}.\mathbf{x}_j\right).$$

$$\tag{1.11}$$

If there are two clusters H_1 and H_2 then the sum in right hand side can be written separately for particles in two clusters

$$C_{\mathbf{k}} = \frac{i}{M} \sum_{j \in H_1} (\mathbf{k}.\mathbf{f}_j) \exp\left(i\mathbf{k}.\mathbf{x}_j\right) + \frac{i}{M} \sum_{j \in H_2} (\mathbf{k}.\mathbf{f}_j) \exp\left(i\mathbf{k}.\mathbf{x}_j\right). \tag{1.12}$$

Representing the force acting on particles in halo '1' due to halo '2' by $\mathbf{f}^{12}$, and the force due to particles within the same halo by $\mathbf{f}^{11}$, we can rewrite the sum[2] as:

$$C_\mathbf{k} = \frac{i}{M} \sum_{j \in H_1} \left(\mathbf{k}.\mathbf{f}_j^{11}\right) \exp\left(i\mathbf{k}.\mathbf{x}_j\right) + \frac{i}{M} \sum_{j \in H_1} \left(\mathbf{k}.\mathbf{f}_j^{12}\right) \exp\left(i\mathbf{k}.\mathbf{x}_j\right)$$
$$+ \frac{i}{M} \sum_{j \in H_2} \left(\mathbf{k}.\mathbf{f}_j^{22}\right) \exp\left(i\mathbf{k}.\mathbf{x}_j\right) + \frac{i}{M} \sum_{j \in H_2} \left(\mathbf{k}.\mathbf{f}_j^{21}\right) \exp\left(i\mathbf{k}.\mathbf{x}_j\right). \quad (1.13)$$

If the wave numbers of interest are such that $|\mathbf{k}.\mathbf{r}| \ll 1$ then we can expand the exponential in each of the terms. Applying Newton's third law, we are left with a simplified expression. If we ignore tidal interaction between the two haloes then we get:

$$C_\mathbf{k} \simeq \frac{i}{M}\mathbf{k}. \left[\sum_{j \in H_1} \mathbf{f}_j^{11} \left(\mathbf{k}.\mathbf{x}_j\right) + \sum_{j \in H_2} \mathbf{f}_j^{22} \left(\mathbf{k}.\mathbf{x}_j\right) \right]$$
$$+ \frac{i}{M}\mathbf{k}.\mathbf{F}^{12} \left(\mathbf{k}. \left(\mathbf{X}_1 - \mathbf{X}_2\right)\right) + \mathcal{O}(k^3). \quad (1.14)$$

Here $\mathbf{F}^{ij}$ is the force due to the jth halo on the ith halo, and $\mathbf{X}_i$ is the centre of mass of the ith halo. It is easy to see that the combination of reduced terms scales inversely with the separation of the two haloes. We can further simplify matters by working in the centre of mass for these haloes. We use centre of mass coordinates $\mathbf{y}_j^i = \mathbf{x}_j - \mathbf{X}_i$ for the jth particle in the ith halo. We will drop the superscript as it is obvious from the context.

$$C_\mathbf{k} = \frac{i}{M}\mathbf{k}. \left[\sum_{j \in H_1} \mathbf{f}_j^{11} \left(\mathbf{k}.\mathbf{y}_j\right) + \sum_{j \in H_2} \mathbf{f}_j^{22} \left(\mathbf{k}.\mathbf{y}_j\right) \right]$$
$$+ \frac{i}{M}\mathbf{k}.\mathbf{F}^{12} \left(\mathbf{k}. \left(\mathbf{X}_1 - \mathbf{X}_2\right)\right) + \mathcal{O}(k^3). \quad (1.15)$$

This follows as the internal forces cancel out for each halo. This is the final expression for $C_\mathbf{k}$, ignoring the tidal interaction terms between the two haloes.

Now we consider the second mode coupling term

$$D_\mathbf{k} = \frac{1}{M} \sum_j m_j \left(\mathbf{k}.\dot{\mathbf{x}}_j\right)^2 \exp\left[i\mathbf{k}.\mathbf{x}_j\right]. \quad (1.16)$$

---

[2] Note that there is no self-interaction implied, each particle experiences force due to all the others.

It is evident that this term will contribute at $\mathcal{O}(k^2)$ or higher orders to the evolution of $\delta_{\mathbf{k}}$. We can again write the sum in two parts, one for each halo:

$$D_{\mathbf{k}} = \frac{1}{M} \sum_j m_j \left(\mathbf{k}.\dot{\mathbf{x}}_j\right)^2 \exp\left[i\mathbf{k}.\mathbf{x}_j\right]$$

$$= \frac{1}{M} \sum_{j \in H_1} m_j \left(\mathbf{k}.\dot{\mathbf{x}}_j\right)^2 + \frac{1}{M} \sum_{j \in H_1} m_j \left(\mathbf{k}.\dot{\mathbf{x}}_j\right)^2 + \mathcal{O}(k^3). \quad (1.17)$$

We now switch to the centre of mass frame for each halo. Let $\mathbf{V}$ be the velocity of the centre of mass and $\dot{\mathbf{y}}_j$ the velocity of the jth particle in this frame:

$$D_{\mathbf{k}} = \frac{1}{M} \sum_{j \in H_1} m_j \left[\mathbf{k}.\left(\dot{\mathbf{y}}_j + \mathbf{V}_1\right)\right]^2 + \frac{1}{M} \sum_{j \in H_1} m_j \left[\mathbf{k}.\left(\dot{\mathbf{y}}_j + \mathbf{V}_2\right)\right]^2 + \mathcal{O}(k^3)$$

$$= \frac{1}{M} \sum_{j \in H_1} m_j \left(\mathbf{k}.\dot{\mathbf{y}}_j\right)^2 + \frac{1}{M} \sum_{j \in H_2} m_j \left(\mathbf{k}.\dot{\mathbf{y}}_j\right)^2$$

$$+ \frac{1}{M} \left[M_1 \left(\mathbf{k}.\mathbf{V}_1\right)^2 + M_2 \left(\mathbf{k}.\mathbf{V}_2\right)^2\right] + \mathcal{O}(k^3). \quad (1.18)$$

The total contribution to mode coupling can be written as:

$$C_{\mathbf{k}} - D_{\mathbf{k}} = \frac{i}{M}\mathbf{k}. \left[\sum_{j \in H_1} \mathbf{f}_j^{11}\left(\mathbf{k}.\mathbf{y}_j\right) + \sum_{j \in H_2} \mathbf{f}_j^{22}\left(\mathbf{k}.\mathbf{y}_j\right)\right]$$

$$+ \frac{1}{M} \sum_{j \in H_1} m_j \left(\mathbf{k}.\dot{\mathbf{y}}_j\right)^2 + \frac{1}{M} \sum_{j \in H_2} m_j \left(\mathbf{k}.\dot{\mathbf{y}}_j\right)^2$$

$$+ \frac{i}{M}\mathbf{k}.\mathbf{F}^{12}\left(\mathbf{k}.\left(\mathbf{X}_1 - \mathbf{X}_2\right)\right) + \frac{1}{M}\left[M_1 \left(\mathbf{k}.\mathbf{V}_1\right)^2 + M_2 \left(\mathbf{k}.\mathbf{V}_2\right)^2\right]$$

$$+ \mathcal{O}(k^3). \quad (1.19)$$

Using the fact that $C_{\mathbf{k}} - D_{\mathbf{k}}$ for a virialized halo, computed in its centre of mass is zero at this order [33], we find:

$$C_{\mathbf{k}} - D_{\mathbf{k}} = \frac{i}{M}\mathbf{k}.\mathbf{F}^{12}\left(\mathbf{k}.\left(\mathbf{X}_1 - \mathbf{X}_2\right)\right)$$

$$+ \frac{1}{M}\left[M_1 \left(\mathbf{k}.\mathbf{V}_1\right)^2 + M_2 \left(\mathbf{k}.\mathbf{V}_2\right)^2\right] + \mathcal{O}(k^3). \quad (1.20)$$

Thus the contribution of interacting haloes to mode coupling leading to influence of small scales on larger scales can be thought of in terms of haloes acting as point masses to the leading order. Here we have ignored the subdominant tidal interactions between the haloes. Both the leading and subleading terms lead to generation of a power spectrum that goes as

k^4 at small k. Subleading terms arise due to tidal interaction and aspherical shapes of haloes [34]. This treatment can be generalized to any number of haloes.

The above discussion has the following obvious implications.

- Approximation of objects such as galaxies or clusters of galaxies by point masses in N-Body simulations does not introduce significant errors in the evolution of large scale structure.
- The small magnitude of the influence of perturbations at small scales on perturbations at larger scales is also explained by this. Clearly, the effect arises from motions of haloes and the dominant effect is due to close pairs of haloes. The time scale over which a close pair of haloes interact and merge to form larger haloes is fairly small. Further, the number of close pairs of haloes is also a very small fraction of all pairs of haloes. Hence the overall effect of small scale motions on larger scales is expected to be small.

1.3.2 *Mode coupling: Effect of large scale perturbations*

We now turn our attention to the effect of perturbations at large scales on perturbations at small scales. Given that the effect of perturbations at small scales on perturbations at much larger scales in known to be small, it is clear that the collapse of perturbations at larger scales is dominated by infall. The question we would like to address is the following: is the collapse of perturbations at a given scale influenced by perturbations at much larger scales?

Formation of highly overdense haloes is often described using the theory of mass function and the excursion set formalism. This formalism is based on the extreme assumption that the collapse of a local overdensity is not influenced by perturbations at larger scales [35, 36]. If this formalism is correct then we should be able to describe the evolution of the spectrum of masses of collapsed haloes for any given model. A comparison with simulations shows that the theory does not fare too badly, even though there are some clear deviations [37]. In the more modern approach, the local overdensity is modeled as a constant overdensity ellipsoid [38], and this matches N-Body simulations much better than the model based on spherical collapse [39]. If the assumptions of the excursion set model are valid then it should be possible to scale out the dependence on the initial conditions as well as the cosmological model. One should, in such a case, obtain the mass function in a universal form. Several authors have tried to test this hypothesis in recent years [40, 41, 42, 43].

I will discuss our attempt [42] in this direction in some detail here. We studied mass functions in a series of models with a different slope for the power spectrum of density fluctuations. We were able to demonstrate, explicitly, that the "universal" form of the mass function does not fit the data for all the models equally well. Indeed, we demonstrated that the parameters of the mass function depend on the slope of the power spectrum of initial density fluctuations. Variation with the slope of the power spectrum was found to be relatively small, smaller than, e.g., the difference between the spherical collapse and ellipsoidal collapse based models. We may interpret these results in the following manner.

- The mass function is not universal, it has a weak dependence on the initial spectrum of density perturbations.
- Departure from universality is small. Our conjecture is that this departure is due to the different tidal influence of the perturbations in different models.
- If verified, this will give us a handle on the influence of perturbations at large scales on perturbations at small scales.

1.4   Conclusions

I have discussed a few issues related to non-linear gravitational clustering in an expanding universe here. I have outlined how the evolution of density perturbations is a strongly non-linear problem. Given the structure of the equations governing gravitational clustering, it is indeed surprising that we are able to provide a nearly universal statistical description for evolution of density perturbations. I have outlined an argument that shows that the perturbations at small scales should not influence collapse of perturbations at larger scales. This is perhaps the reason for the existence of a nearly universal mapping from the initial correlation function to the final correlation functions.

I also discussed the converse problem of the influence of perturbations at large scales on collapse of a given overdense region. A comparison of simulations with the models shows that we can ignore the influence of large scales as a first approximation and that the major influence may be due to the tidal force of the large scale distribution on the collapse of perturbation.

Together, these imply that the most significant process in gravitational clustering is infall on to overdensities and the details of mode coupling essentially provide (relatively small) corrections.

Acknowledgments

I would like to thank Paddy for introducing me to this area of physics, and for numerous discussions that we have had on this and many other problems in physics.

References

[1] Hoyle F., 1953, ApJ, 118, 513
[2] Silk J., 1977, ApJ, 211, 638
[3] Rees M. J., Ostriker J. P., 1977, MNRAS, 179, 541
[4] Binney J., 1977, ApJ, 215, 483
[5] Peebles P. J. E., 1980, Large Scale Structure in the Universe, Princeton University Press.
[6] Padmanabhan, T. 1993, Structure Formation in the Universe, by T. Padmanabhan, pp. 499. ISBN 0521424860. Cambridge, UK: Cambridge University Press
[7] Peacock J. A., 1999, Cosmological Physics, Cambridge University Press.
[8] Bernardeau F., Colombi S., Gaztañaga E., Scoccimarro R., 2002, PhR, 367, 1
[9] Padmanabhan T., 2002, A Course in Theoretical Astrophysics vol.3, Cambridge University Press.
[10] Heath, D. J. 1977, MNRAS, 179, 351
[11] Zel'Dovich Y. B., 1970, A&A, 5, 84
[12] Shandarin, S. F., & Zeldovich, Y. B. 1989, Reviews of Modern Physics, 61, 185
[13] Coles, P., Melott, A. L., & Shandarin, S. F. 1993, MNRAS, 260, 765
[14] Bond, J. R., & Myers, S. T. 1996, ApJS, 103, 1
[15] Taylor, A. N., & Hamilton, A. J. S. 1996, MNRAS, 282, 767
[16] Padmanabhan, T. 1999, arXiv:astro-ph/9911374
[17] Padmanabhan, T., & Subramanian, K. 1993, ApJ, 410, 482
[18] Brainerd T. G., Scherrer R. J., Villumsen J. V., 1993, ApJ, 418, 570
[19] Bagla J. S., Padmanabhan T., 1994, MNRAS, 266, 227

[20] Efstathiou G., Frenk C. S., White S. D. M., Davis M., 1988, MNRAS, 235, 715

[21] Hamilton A. J. S., Kumar P., Lu E., Matthews A., 1991, ApJ, 374, L1

[22] Nityananda R., Padmanabhan T., 1994, MNRAS, 271, 976

[23] Padmanabhan T., 1996, MNRAS, 278, L29

[24] Fillmore J. A., Goldreich P., 1984, ApJ, 281, 1

[25] Smith R. E., et al., 2003, MNRAS, 341, 1311

[26] Bagla J. S., Padmanabhan T., 1997, MNRAS, 286, 1023

[27] Crocce M., Scoccimarro R., 2006, PhRvD, 73, 063519

[28] McDonald P., 2007, PhRvD, 75, 043514

[29] Little B., Weinberg D. H., Park C., 1991, MNRAS, 253, 295

[30] Klypin A. A., Melott A. L., 1992, ApJ, 399, 397

[31] Bagla J. S., Prasad J., Ray S., 2005, MNRAS, 360, 194

[32] Bagla J. S., Prasad J., 2009, MNRAS, 393, 607

[33] Peebles P. J. E., 1974, A&A, 32, 391

[34] Bagla J. S., 2010, In Preparation.

[35] Press, W. H., & Schechter, P. 1974, ApJ, 187, 425

[36] Bond, J. R., Cole, S., Efstathiou, G., & Kaiser, N. 1991, ApJ, 379, 440

[37] Lacey, C., & Cole, S. 1994, MNRAS, 271, 676

[38] Sheth, R. K., & Tormen, G. 1999, MNRAS, 308, 119

[39] Gunn, J. E., & Gott, J. R., III 1972, ApJ, 176, 1

[40] Lukić, Z., Heitmann, K., Habib, S., Bashinsky, S., & Ricker, P. M. 2007, ApJ, 671, 1160

[41] Tinker, J., Kravtsov, A. V., Klypin, A., Abazajian, K., Warren, M., Yepes, G., Gottlöber, S., & Holz, D. E. 2008, ApJ, 688, 709

[42] Bagla, J. S., Khandai, N., & Kulkarni, G. 2009, arXiv:0908.2702

[43] Courtin, J., Rasera, Y., Alimi, J. -., Corasaniti, P. -., Boucher, V., & Fuzfa, A. 2010, arXiv:1001.3425

Chapter 2

Dark ages and cosmic reionization

Tirthankar Roy Choudhury

Harish-Chandra Research Institute,
Chhatnag Road, Jhusi, Allahabad 211 019, India.
E-mail: *tirth@hri.res.in*

Abstract: About 300,000 years after the Big Bang, the protons and the electrons combined for the first time in the Universe to form hydrogen (and helium) atoms, which is known as the recombination epoch. Following that, the Universe entered a phase called the "dark ages" where no significant radiation sources existed. The dark ages ended once the first structures collapsed and luminous sources like stars and accreting black holes started forming. The radiation from these sources then ionized hydrogen atoms in the surrounding medium, a process known as "reionization". Reionization is thus the second major change in the ionization state of hydrogen (and helium) in the Universe (the first being the recombination).

The study of dark ages and cosmic reionization has acquired increasing significance over the last few years because of various reasons. On the observational front, we now have good quality data of different types at high redshifts (quasar absorption spectra, radiation backgrounds at different frequencies, number counts of galaxies, cosmic microwave background polarization, Lyα emitters and so on). Theoretically, the importance of the reionization lies in its close coupling with the formation of first cosmic structures, and there have been numerous progresses in modeling the process. In this article, we introduce the basic concepts involving the formation of first structures and evolution of the ionization history of the Universe. We also discuss the possibility of constraining the reionization history by matching theoretical models with observations.

2.1   Introduction

Study of reionization mostly concerns with the ionization and thermal history of the baryons (hydrogen and helium) in our Universe [36, 2, 14, 19]. Within the framework of hot Big Bang model, hydrogen formed for the first time when the age of the Universe was about 3×10^5 years, its size being one-thousandth of the present (corresponding to a scale factor $a \approx 0.001$ and a redshift $z = 1/a - 1 \approx 1000$). The epoch at which the protons and the electrons combined for the first time to form hydrogen atoms is known as the recombination epoch and is well-probed by the Cosmic Microwave Background Radiation (CMBR).

Right after the recombination epoch, the Universe entered a phase called the "dark ages" where no significant radiation sources existed. The hydrogen remained largely neutral at this phase. The small inhomogeneities in the dark matter density field which were present during the recombination epoch started growing via gravitational instability giving rise to highly nonlinear structures like the collapsed haloes. It should, however, be kept in mind that most of the baryons at high redshifts do not reside within these haloes, they rather reside as diffuse gas within the intergalactic space which is known as the intergalactic medium (IGM) [47, 45].

The collapsed haloes form potential wells whose depth depend on their mass and the baryons (i.e, hydrogen) then "fall" in these wells. If the mass of the halo is high enough (i.e., the potential well is deep enough), the gas will be able to dissipate its energy, cool via atomic or molecular transitions and fragment within the halo. This produces conditions appropriate for condensation of gas and forming stars and galaxies. Once these luminous objects form, the era of dark ages can be thought of being over.

The first population of luminous stars and galaxies can generate ultraviolet (UV) radiation through nuclear reactions. In addition to the galaxies, perhaps an early population of accreting black holes (quasars) also generated some amount UV radiation. The UV radiation contains photons with energies > 13.6 eV which are then able to ionize hydrogen atoms in the surrounding medium, a process known as "reionization". Reionization is thus the second major change in the ionization state of hydrogen (and helium) in the Universe (the first being the recombination).

As per our current understanding, reionization started around the time when first structures formed, which is currently believed to be around $z \approx 20 - 30$. In the simplest picture, each source first produced an ionized region around it; these regions then overlapped and percolated into the

IGM. This era is usually called the "pre-overlap" phase. The process of overlapping seemed to be completed around $z \approx 6-8$ at which point the neutral hydrogen fraction fell to values lower than 10^{-4}. Following that a never-ending "post-reionization" (or "post-overlap") phase started which implies that the Universe is largely ionized at present epoch. Reionization by UV radiation is also accompanied by heating: electron which are released by photoionization will deposit an extra energy equivalent to $h_p\nu - 13.6$ eV to the IGM, where ν is the frequency of the ionizing photon and h_p is the Planck constant. This reheating of the IGM can expel the gas and/or suppress cooling in the low mass haloes – thus, there is a considerable reduction in the cosmic star formation right after reionization.

The process of reionization is of immense importance in the study of structure formation since, on one hand, it is a direct consequence of the formation of first structures and luminous sources while, on the other, it affects subsequent structure formation. Observationally, the reionization era represents a phase of the Universe which is yet to be probed; the earlier phases ($z \approx 1000$) are probed by the CMBR while the post-reionization phase ($z < 6$) is probed by various observations based on galaxies, clusters, quasars and other sources. In addition to the importance outlined above, the study of dark ages and cosmic reionization has acquired increasing significance over the last few years because of the availability of good quality data in different areas.

In this article, we will introduce various concepts which go into modelling reionization. The main aim would be to systematically discuss the set of equations which are crucial in understanding the process highlighting the major physical processes and assumptions. We shall also highlight the relevant observational probes at appropriate places. In Section 2.2, we shall give a pedagogic introduction to the basic theoretical formalism for studying reionization and IGM in different phases of evolution. Section 2.3 would be devoted to discussing detailed modelling of reionization using the formalism developed. We shall illustrate on how to constrain the models by comparing with a wide variety of available data sets. In Section 2.4, we shall briefly discuss the current numerical simulations and observations related to reionization. We shall also briefly highlight what to expect in this field in near future. More detailed discussions on this topic and detailed reference lists can be found elsewhere [12].

2.2   Theoretical formalism

In this section, we discuss the basic theoretical formalism required for modelling reionization of the IGM. The main aim here would be to highlight the physical processes which are crucial in understanding reionization and comparing with observations. In what follows, we shall assume that the IGM consists only of hydrogen and neglect the presence of helium. It is straightforward to include helium into the formalism.

Essentially, in presence of a ionizing radiation, the evolution of the mean neutral hydrogen density $n_{\rm HI}$ [1] is given by

$$\dot{n}_{\rm HI} = -3H(t)n_{\rm HI} - \Gamma_{\rm HI}n_{\rm HI} + \mathcal{C}\alpha(T)n_{\rm HII}n_e \qquad (2.1)$$

where overdots denote the total time derivative d/dt, $H \equiv \dot{a}/a$ is the Hubble parameter, $\Gamma_{\rm HI}$ is the photoionization rate per hydrogen atom, $\alpha(T)$ is the recombination rate coefficient and n_e represents the mean electron density. The first term in the right hand side of equation (2.1) corresponds to the dilution in the density because of cosmic expansion, the second term corresponds to photoionization by the ionizing flux and the third term corresponds to recombination of protons and free electrons into neutral hydrogen. The quantity $\mathcal{C}$ is called the clumping factor and is defined as

$$\mathcal{C} \equiv \frac{\langle n_{\rm HII}n_e \rangle}{\langle n_{\rm HII} \rangle \langle n_e \rangle} = \frac{\langle n_H^2 \rangle}{\langle n_H \rangle^2} \qquad (2.2)$$

where the last equality holds for the case when the IGM contains only hydrogen (i.e., no helium) and is highly ionized, i.e, $n_e = n_{\rm HII} \approx n_H$. The clumping factor takes into account the fact that the recombination rate in an inhomogeneous (clumpy) IGM is higher than a medium of uniform density.

The ionization equation is usually supplemented by the evolution of the IGM temperature T, which is given by

$$\dot{E}_{\rm kin} = -2H(t)E_{\rm kin} + \Lambda \qquad (2.3)$$

where $E_{\rm kin} = 3k_B T n_H$ is the kinetic energy of the gas and Λ is the net heating rate including all possible heating and cooling processes. The first term on the right hand side takes into account the adiabatic cooling of the gas because of cosmic expansion.

---

[1] In astrophysical notation, HI stands for neutral hydrogen while HII denotes ionized hydrogen (proton).

2.2.1 Cosmological radiation transfer

The equation of radiation transfer, which describes propagation of radiation flux through a medium, is written as an evolution equation for the specific intensity of radiation $I_\nu \equiv I(t, \mathbf{x}, \nu, \hat{\mathbf{n}})$ which has dimensions of the energy per unit time per unit area per unit solid angle per frequency range. It is a function of time and space coordinates $(t, \mathbf{x})$, the frequency of radiation ν and the direction of propagation $\hat{\mathbf{n}}$. The radiation transfer equation in a cosmological scenario has the form

$$\frac{\partial I_\nu}{\partial t} + \frac{c}{a(t)} \hat{\mathbf{n}} \cdot \nabla_{\mathbf{x}} I_\nu - H(t)\nu \frac{\partial I_\nu}{\partial \nu} + 3H(t)I_\nu$$
$$= -c\kappa_\nu I_\nu + \frac{c}{4\pi} \epsilon_\nu, \tag{2.4}$$

where κ_ν is the absorption coefficient and ϵ_ν is the emissivity. The above equation is essentially the Boltzmann equation for photons with I_ν being directly proportional to the phase space distribution function [45]. The terms on the left hand side of equation (2.4) add up to the total time derivative of I_ν; in particular, the third term corresponds to dilution of the intensity and the fourth term accounts for shift of frequency $\nu \propto a^{-1}$ because of cosmic expansion. The effect of scattering (which is much rarer than absorption in the IGM) can, in principle, be included in the κ_ν term if required. If the medium contains absorbers with number density n_{abs} each having a cross-section σ_ν, the absorption coefficient is given by $\kappa_\nu = n_{\text{abs}}\sigma_\nu$. The mean free path of photons in the medium is given by $\lambda_\nu(t) \equiv \kappa_\nu^{-1}(t)$.

We define the mean specific intensity by averaging I_ν over a large volume and over all directions

$$J_\nu(t) \equiv \int_V \frac{d^3x}{V} \int \frac{d\Omega}{4\pi} I_\nu(t, \mathbf{x}, \hat{\mathbf{n}}) \tag{2.5}$$

Then the spatially and angular-averaged radiation transfer equation becomes [48]

$$\dot{j}_\nu \equiv \frac{\partial J_\nu}{\partial t} - H(t)\nu \frac{\partial J_\nu}{\partial \nu} = -3H(t)J_\nu - c\kappa_\nu J_\nu + \frac{c}{4\pi} \epsilon_\nu \tag{2.6}$$

where the coefficients κ_ν and ϵ_ν are now assumed to be averaged over the large volume. The quantity J_ν is essentially the energy per unit time per unit area per frequency interval per solid angle.

The integral solution of the above equation along a line of sight can be written as [29]

$$J_\nu(t) = \frac{c}{4\pi} \int_0^t dt' \epsilon_{\nu'}(t') \left[\frac{a^3(t')}{a^3(t)} \right] e^{-\tau(t, t'; \nu)}. \tag{2.7}$$

where $\nu' = \nu a(t)/a(t')$, $\nu'' = \nu a(t)/a(t'')$ and

$$\tau(t,t';\nu) \equiv c \int_{t'}^{t} dt'' \kappa_{\nu''}(t'') = c \int_{t'}^{t} \frac{dt''}{\lambda_{\nu''}(t'')} \tag{2.8}$$

is the optical depth along the line of sight from t' to $t > t'$ Clearly, the intensity at a given epoch is proportional to the integrated emissivity with an exponential attenuation due to absorption in the medium. The intensity attenuates by $1/e$ when the radiation travels a distance equal to the mean free path.

The absorption is "local" when the mean free path of photons is much smaller than the horizon size of the Universe, i.e., $\lambda_\nu(t) \ll c/H(t)$. In addition, if we also assume that the emissivity ϵ_ν does not evolve significantly over the small time interval λ/c, then the specific intensity is related to the emissivity through a simple form [38, 43]

$$J_\nu(t) \approx \frac{\epsilon_\nu(t)\lambda_\nu(t)}{4\pi} = \frac{\epsilon_\nu(t)}{4\pi\kappa_\nu(t)} \tag{2.9}$$

Note that in the case of local absorption, $\dot{J}, HJ \ll c\epsilon$. In this approximation, the background intensity depends only on the instantaneous value of the emissivity (and not its history) because all the photons are absorbed shortly after being emitted (unless the sources evolve synchronously over a timescale much shorter than the Hubble time). We shall discuss later in Section 2.2.2 that this is a useful approximation for the IGM for redshifts $z \gtrsim 3$.

2.2.2 *Post-reionization epoch*

Let is first study the radiation transfer in the post-reionization epoch. Compared to the pre-overlap era, this epoch is much easier to study because the IGM can be treated as a highly ionized single-phase medium (whereas during the pre-overlap era, one is looking into two distinct phases – ionized and neutral). The optical depth can be written as

$$\tau(z,z';\nu) = c \int_{z}^{z'} \frac{dz''}{(1+z'')H(z'')} n_{\rm HI}(z'') \, \sigma_{\rm abs}(\nu'') \tag{2.10}$$

where $\sigma_{\rm abs}(\nu)$ is the total absorption cross section of neutral hydrogen and we have changed the time coordinate to the redshift z. Various processes can, in principle, contribute to $\sigma_{\rm abs}(\nu)$, most dominant being the resonant Lyman series absorption corresponding to excitation of hydrogen atoms from the ground state to higher ones (1s $\rightarrow$ np) and the continuum absorption of photons above the ionization threshold via photoionization process. Let us discuss them in the following:

2.2.2.1 *Resonant Lyman series absorption*

The Lyman series absorption arises from the electronic excitation of neutral hydrogen atoms from the 1s ground state to higher ones. The most dominant of these are the Lyα (1s $\to$ 2p, rest wavelength $\lambda_\alpha \approx 1216$ Å) and Lyβ (1s $\to$ 3p, $\lambda_\beta \approx 1206$ Å) transitions, and hence they are the most relevant ones as far as observations are concerned. For simplicity, we shall present results for the Lyα absorption only, the others can be calculated in identical manner. The Lyα absorption cross section is given by

$$\sigma_{\mathrm{abs}}(\nu) = \sigma_\alpha \, V \left(\frac{\nu}{\nu_\alpha} - 1 \right) \tag{2.11}$$

where $\nu_\alpha = c/\lambda_\alpha$ is the resonant frequency of transition, $\sigma_\alpha = 4.45 \times 10^{-18} \mathrm{cm}^2$ is the cross section at ν_{alpha} and V is a function which determines the profile of the absorption line. It is called the Voigt profile function and is a convolution of the Lorentzian shape for the natural broadening and the Gaussian shape for the thermal broadening. For the purpose of this article, it is sufficient to note that V is a sharply peaked function about $\nu/\nu_\alpha = 1$; for most our discussion, we shall take it to be a Dirac-delta function $V(\nu/\nu_\alpha - 1) = \delta_D(\nu/\nu_\alpha - 1)$.

The optical depth between the redshifts z and z' is then given by

$$\tau(z, z'; \nu) = \tau(z_\alpha) = \sigma_\alpha \frac{c}{H(z_\alpha)} n_{\mathrm{HI}}(z_\alpha);$$

$$1 + z_\alpha = \frac{\nu_\alpha}{\nu}(1 + z) \tag{2.12}$$

If we put this into equation (2.7), we see that the Lyα absorption at a redshift z_α reduces the specific intensity observed at z at a frequency $\nu_\alpha(1 + z)/(1 + z_\alpha)$ by a factor $e^{-\tau(z_\alpha)}$. The value of $\tau(z_\alpha)$ along a given line of sight would depend upon the distribution of $n_{\mathrm{HI}}(z_\alpha)$. However, we would mostly be interested in the mean value of specific intensity averaged over a number of lines of sight. The corresponding reduction can be described by a line-of-sight-averaged optical depth

$$e^{-\tau_{\mathrm{eff}}(z_\alpha)} \equiv \langle e^{-\tau(z_\alpha)} \rangle_{\mathrm{LOS}} \tag{2.13}$$

where $\langle \ \rangle_{\mathrm{LOS}}$ denotes averaging over lines of sight. The quantity τ_{eff} is usually known as the "effective optical depth".

Theoretically, the value of τ_{eff} can be calculated if we know the distribution of optical depth $P(\tau)$ [which can be calculated from the neutral hydrogen distribution $P(n_{\mathrm{HI}})$]:

$$e^{-\tau_{\mathrm{eff}}(z)} = \int_0^\infty d\tau \, P(\tau; z) \, e^{-\tau} \tag{2.14}$$

Of course, one requires detailed understanding of the evolution of the baryonic density field to model the distribution $P(n_{HI})$, which we shall avoid discussing here. However, we can still make some inference assuming the distribution is uniform, i.e., $\tau_{eff}(z) = c\sigma_\alpha n_{HI}(z)/H(z)$. If we define the neutral hydrogen fraction to be $x_{HI} \equiv n_{HI}/n_H$, then we can calculate $\tau_{eff} \propto x_{HI}$ given a set of cosmological parameters (which would uniquely determine $H(z)$ and n_H).

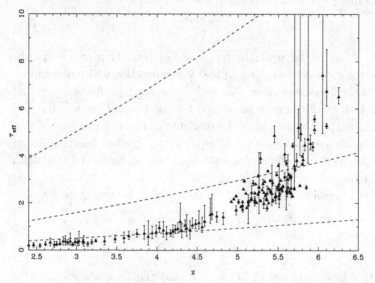

Fig. 2.1   The effective optical depth of Lyα absorption as function of redshift z. The points with errorbars represent the observational data. The dashed curves, from top to bottom, represent the predictions for a uniform IGM with neutral hydrogen fraction $x_{HI} = (3, 1, 0.3) \times 10^{-5}$, respectively.

Observationally, τ_{eff} can be determined by looking at the spectra of bright sources like quasars at high redshifts. These spectra show a series of absorption features at frequencies larger than the Lyα frequency in the quasar rest frame. Since one has a good knowledge of the unabsorbed quasar spectra (from looking at nearby quasars and also having some understanding about the physical processes), one can calculate the amount of absorption happening because of the intervening IGM between the quasar and the observer; this absorption, averaged over numerous lines of sight, is essentially the quantity $e^{-\tau_{eff}}$. This has been done to quite high redshifts

$z \sim 6.5$ and the values of τ_{eff} observed are shown as points with errorbars [61, 20] in Figure 2.1.

To understand what these values imply, we have plotted with dashed lines the calculated value of τ_{eff} for a uniform IGM assuming three values of $x_{\text{HI}} = (3, 1, 0.3) \times 10^{-5}$ from top to bottom. This immediately tells us that the fraction of neutral hydrogen has to be $\sim 10^{-5}$ in order to reproduce the observed values of τ_{eff}. In fact, had x_{HI} been slightly (say $\sim 10^{-4}$) higher, one would have obtained τ_{eff} much higher than unity ($\sim 10 - 100$) and hence the flux from the quasar would be completely absorbed. If that were the case, it would show up as a absorption "trough" at frequencies larger than the rest frame Lyα frequency. In reality, a considerable amount of transmitted flux is found at these frequencies alongwith a series of absorption features arising from the Lyα transition of residual neutral hydrogen. These absorption signatures are known as the "Lyα" forest and are powerful probes of the neutral hydrogen distribution in the IGM at $z < 6$ [52].

The absence of a absorption trough is a direct proof of the fact that hydrogen is completely reionized in the diffuse IGM at redshifts $z \lesssim 6$. This is known as the Gunn-Peterson effect [28]. Note that the actual inferred value of x_{HI} might be slightly different if one models with an appropriate density distribution, however, the basic conclusion remains unchanged.

We can also see from Figure 2.1 that for quasars at redshifts $z \gtrsim 6$, the observed value of $\tau_{\text{eff}} \gtrsim 5$; this would imply an attenuation $\gtrsim 0.01$ and hence one actually observes absorption troughs as predicted by Gunn-Peterson effect [20, 68]. Unfortunately, finding such troughs does not necessarily imply that the IGM is highly neutral as even a $x_{\text{HI}} \sim 10^{-4}$ could be sufficient to absorb all the flux. However, one can use much detailed modelling to improve the constraint, which we shall discuss later in Section 2.3.2.

The values of $\tau_{\text{eff}} \lesssim 1$ at $z < 4$ means that the diffuse IGM is highly transparent (also called optically thin) to Lyα photons. Only about $\sim 10\%$ of the Lyα photons are absorbed, mostly within the high density regions. These high density systems are often modelled as a set of discrete absorbers of some size. If we consider an absorber having neutral hydrogen density n_{HI} and a size $L \ll c/H(z)$ along the line of sight at a redshift z_{abs}, the optical depth is given by

$$\tau(z, z'; \nu) = N_{\text{HI}} \, \sigma_{\text{abs}}(\nu_{\text{abs}}) \tag{2.15}$$

where $\nu_{\text{abs}} = \nu(1 + z_{\text{abs}})/(1 + z)$ and $N_{\text{HI}} \equiv n_{\text{HI}} L$ is the column density of neutral hydrogen within the absorber. Hence, each absorber reduces the specific intensity by a factor $e^{-N_{\text{HI}}\sigma_{\text{abs}}(\nu_{\text{abs}})}$. If we assume that the absorbers are Poisson-distributed, then it is straightforward to show that the effective optical depth is given by [46]

$$\tau_{\text{eff}}(z, z'; \nu) =$$
$$\int_{z}^{z'} dz'' \int_{0}^{\infty} dN_{\text{HI}} \frac{\partial^2 N}{\partial z'' \partial N_{\text{HI}}} [1 - e^{-N_{\text{HI}} \sigma_{\text{abs}}(\nu'')}] \qquad (2.16)$$

where $[\partial^2 N / \partial z \partial N_{\text{HI}}] \, dz \, dN_{\text{HI}}$ is the number of absorbers within $(z, z+dz)$ having column densities in the range $(N_{\text{HI}}, N_{\text{HI}} + dN_{\text{HI}})$.

In case of Lyα resonant absorption, we can use the cross section in (2.11) to calculate τ_{eff}. Since V is a function which is sharply peaked around $\nu/\nu_{\alpha} = 1$, we can approximate the above integral as

$$\tau_{\text{eff}}(z, z'; \nu) = \frac{1 + z_{\alpha}}{\lambda_{\alpha}} \int_{0}^{\infty} dN_{\text{HI}} \frac{\partial^2 N}{\partial z_{\alpha} \partial N_{\text{HI}}} W_{\alpha}(N_{\text{HI}}) \qquad (2.17)$$

where

$$W_{\alpha}(N_{\text{HI}}) \equiv \int d\lambda'' \left[1 - e^{-N_{\text{HI}} \sigma_{\alpha} V(\lambda_{\alpha}/\lambda'' - 1)} \right] \qquad (2.18)$$

is called the "equivalent width" of the absorber.

2.2.2.2 *Continuum absorption*

In case of continuum absorption of radiation by photoionization, the cross section is given by

$$\sigma_{\text{abs}}(\nu) = \sigma_{\text{HI}}(\nu) \Theta(\nu - \nu_{\text{HI}}) \qquad (2.19)$$

where $\sigma_{\text{HI}}(\nu)$ is the photoionization cross section and Θ is the Heaviside step function taking into account that only photons with frequencies $\nu > \nu_{\text{HI}} = 13.6$ eV$/h_p$ would be absorbed by the photoionization process. The exact form of $\sigma_{\text{HI}}(\nu)$ is rather complicated, however one can approximate it by a power-law of the form $\sigma_{\text{HI}}(\nu) = \sigma_0 (\nu/\nu_{\text{HI}})^{-3}$ where $\sigma_0 = 6.3 \times 10^{-18}$cm^2.

Since $\sigma_0 \sim \sigma_{\alpha}$, one can show that the absorption due to a diffuse IGM in this case too is negligibly small. The only significant absorption can be seen in very high density regions which have a large fraction of their hydrogen in neutral form. In that case, we can use the relations obtained for a set of Poisson-distributed absorbers in the vase of resonant transition. We essentially have an optical depth of the form (2.17), and the corresponding mean free path of ionizing photons due to these discrete absorbers is found to be [38]

$$\lambda_{\nu}(z) = \frac{c}{H(z)(1 + z)}$$
$$\times \left[\int_{0}^{\infty} dN_{\text{HI}} \frac{\partial^2 N}{\partial z \partial N_{\text{HI}}} [1 - e^{-N_{\text{HI}} \sigma_{\text{HI}}(\nu)}] \right]^{-1} \qquad (2.20)$$

At this point, let us introduce the concept of Lyman-limit systems which have column densities $N_{HI} > \sigma_0^{-1} = 1.6 \times 10^{17} \text{cm}^{-2}$; these absorbers contribute an optical depth of unity to the ionizing photons. The average distance between these systems is given by

$$\lambda_{LLS} = \frac{c}{H(z)(1+z)} \left[\frac{dN_{LLS}}{dz} \right]^{-1} \tag{2.21}$$

where dN_{LLS}/dz is the redshift distribution of the Lyman-limit systems

$$\frac{dN_{LLS}}{dz} = \int_{1/\sigma_{HI}(\nu_{HI})}^{\infty} dN_{HI} \frac{\partial^2 N}{\partial z \partial N_{HI}} \tag{2.22}$$

For the observed distribution $\partial^2 N/\partial z \partial N_{HI} \propto N_{HI}^{-1.5}$ [49], one can show from equations (2.20) and (2.21) that the mean free path is related to the distance and redshift distribution Lyman-limit systems as

$$\lambda_{\nu_{HI}} = \frac{\lambda_{LLS}}{\sqrt{\pi}} = \frac{c}{\sqrt{\pi}H(z)(1+z) \, dN_{LLS}/dz}. \tag{2.23}$$

The redshift distribution of Lyman-limit systems dN_{LLS}/dz is a quantity which has been measured for $2 < z < 4.5$ by observations of quasar absorption spectra. Though the observational constraints are poor, one can still obtain a value $dN_{LLS}/dz \approx 0.3(1+z)^{-1.55}$ [65], which in turn gives the mean free path as $\lambda_{\nu_{HI}}/[c/H(z)] \approx 0.1[(1+z)/4]^{-2.55}$ Hence the mean free path of ionizing photons is much smaller than the horizon size for $z > 3$, which implies that we can use the local absorption approximation at these redshifts.

We can summarise the main results of this section as: the post-reionization epoch is characterized by a highly ionized IGM as observed by the quasar absorption spectra. The IGM is largely transparent to ionizing photons at these redshifts. However, there exist regions with high column densities ($N_{HI} > 10^{17} \text{cm}^{-2}$) which are optically thick to the ionizing radiation; these regions determine the photon mean free path. We shall see later how to use this information to obtain an improved model of the IGM.

2.2.3 *Pre-overlap epoch*

We now turn our attention towards the IGM in the pre-overlap era. In this era, the overlap of individual ionized regions is not complete and hence the IGM is partially ionized. So the radiation transfer equation has to be modified to account for the multi-phase nature of the IGM.

Let us define the volume filling factor of ionized regions to be Q_{HII}; this is the fraction of volume that is ionized and reionization is said to be complete when $Q_{HII} = 1$. Next, note that the number density of photons present in the background flux is

$$n_J(t) = \frac{4\pi}{c} \int_{\nu_{HI}}^{\infty} d\nu \frac{J_\nu}{h_p \nu} \tag{2.24}$$

Since there is no ionizing flux within the neutral regions (otherwise they would not remain neutral), the photoionization rate per hydrogen atom *within the ionized (HII) regions* is

$$\Gamma_{HI}^{II} = \frac{1}{Q_{HII}} 4\pi \int_{\nu_{HI}}^{\infty} d\nu \frac{J_\nu}{h_p \nu} \sigma_{HI}(\nu) \tag{2.25}$$

where the factor Q_{HII}^{-1} accounts for the fact that the radiation is limited to a fraction of the total volume. The emission rate of ionizing photons per unit volume from sources of emissivity ϵ_ν is

$$\dot{n}_{ph} = \int_{\nu_{HI}}^{\infty} d\nu \frac{\epsilon_\nu}{h_p \nu} \tag{2.26}$$

Then the equation of radiation transfer becomes [43, 11]

$$\dot{n}_J = -3H(t)n_J - H(t) \frac{4\pi}{c} \frac{J_{\nu_{HI}}}{h_p} + \dot{n}_{ph}$$
$$- n_{HI}^{II} Q_{HII} \Gamma_{HI}^{II} - n_{HII}^{II} \frac{dQ_{HII}}{dt} \tag{2.27}$$

where n_{HI}^{II} and n_{HII}^{II} are the number densities of neutral and ionized hydrogen within the HII regions, respectively. The first term in the right hand side of equation (2.27) corresponds to the dilution in density due to cosmic expansion while the second term accounts for the loss of ionizing radiation because of a photon being redshifted below the ionization edge of hydrogen ν_{HI}. The third term is essentially the source of ionizing photons. The fourth term accounts for the loss of photons in ionizing the residual neutral hydrogen within the ionized regions. The fifth term, which is only relevant for the pre-overlap stages, accounts for the photons which ionize hydrogen for the first time and hence increase the filling factor Q_{HII}. For $Q_{HII} = 1$, equation (2.27) reduces to that for the post-reionization phase.

If we now assume that the photons are absorbed locally, then $\dot{J}, HJ \ll c\epsilon$ and J_ν is essentially given by equation (2.9). We can then ignore terms containing J and n_J in equation (2.27). This gives an equation describing the evolution of the filling factor Q_{HII}

$$\frac{dQ_{HII}}{dt} = \frac{\dot{n}_{ph}}{n_{HII}^{II}} - Q_{HII} \Gamma_{HI}^{II} \tag{2.28}$$

If we further assume photoionization equilibrium within the ionized region $\mathrm{d}(n_{\mathrm{HI}}^{II}a^3)/\mathrm{d}t \to 0$, then we have from equation (2.1) $n_{\mathrm{HI}}^{II}\Gamma_{\mathrm{HI}}^{II} = \mathcal{C}\alpha(T)n_{\mathrm{HII}}^{II}n_e^{II}$ and the evolution of Q_{HII} can be written in the form [38]

$$\frac{\mathrm{d}Q_{\mathrm{HII}}}{\mathrm{d}t} = \frac{\dot{n}_{\mathrm{ph}}}{n_{\mathrm{HII}}^{II}} - Q_{\mathrm{HII}}\,\mathcal{C}\alpha(T)n_e^{II} \tag{2.29}$$

In this description, reionization is complete when $Q_{\mathrm{HII}} = 1$ and equation (2.29) cannot be evolved further on. Clearly the assumptions of local absorption and photoionization equilibrium (both of which are reasonably accurate) has given us an equation which can be solved once we have a model for estimating ϵ_ν and $\mathcal{C}$. Of course, there is a dependence of the recombination rate coefficient $\alpha(T)$ on temperature, however that dependence is often ignored while studying the volume filling factor. In case one is interested in temperature evolution, one has to solve equation (2.3) taking into account all the heating and cooling processes in the IGM, in particular, the photoheating by ionizing photons whose rate is given by

$$\Gamma_{\mathrm{ph,HI}} = 4\pi \int_{\nu_{\mathrm{HI}}}^{\infty} \mathrm{d}\nu \frac{J_\nu}{h_p\nu} h_p(\nu - \nu_{\mathrm{HI}})\sigma_{\mathrm{HI}}(\nu). \tag{2.30}$$

2.2.4 *Reionization of the inhomogeneous IGM*

The description of reionization in the previous section is not adequate as it does not take into account the inhomogeneities in the IGM appropriately (except for a clumping factor $\mathcal{C}$ in the effective recombination rate). To see this, consider the post-reionization phase where we know from observations that there exist regions of high density which are neutral; these regions are being gradually ionized and hence one would ideally like to write an equation similar to (2.29) for studying the post-reionization phase. Since the ionization state depends on the density, one should have to account for the density distribution of the IGM.

In order to proceed, first note that the volume filling factor may not be the appropriate quantity to study for evolution of reionization because most of the photons are consumed in regions with high densities (which might be occupying a small fraction of volume). In other words, if we neglect recombination for the moment, we have from equation (2.29) that the volume filling factor $Q_{\mathrm{HII}} = \int \mathrm{d}t\, \dot{n}_{\mathrm{ph}}/n_H = n_{\mathrm{ph}}/n_H$; however, in reality the photon to hydrogen ratio should be equal to the ionized mass fraction F_{HII}^M, i.e., $n_{\mathrm{ph}}/n_H = F_{\mathrm{HII}}^M$. Hence, we must replace the volume filling factor by the mass filling factor in the description of the previous section, in particular equation (2.29) should have the form

$$\frac{dF_{\text{HII}}^M}{dt} = \frac{\dot{n}_{\text{ph}}}{n_{\text{HII}}^{\text{II}}} - F_{\text{HII}}^M \, \mathcal{C}\alpha(T)n_e^{\text{II}} \tag{2.31}$$

One can relate F_{HII}^M to the IGM density distribution by using the fact that regions of lower densities will be ionized first, and high-density regions will remain neutral for a longer time. The main reason for this is that the recombination rate (which is $\propto n_H^2$) is higher in high-density regions where dense gas becomes neutral very quickly. If we assume that hydrogen in all regions with overdensities $\Delta < \Delta_{\text{HII}}$ is ionized while the rest is neutral, then the mass ionized fraction is clearly [43]

$$F_{\text{HII}}^M \equiv F^M(\Delta_{\text{HII}}) = \int_0^{\Delta_{\text{HII}}} d\Delta \, \Delta \, P(\Delta) \tag{2.32}$$

where $P(\Delta)$ is the (volume-weighted) density distribution of the IGM. The term describing the effective recombination rate gets contribution only from the low density regions (high density neutral regions do not contribute) and is then given by

$$\alpha(T)n_e^{\text{II}} \int_0^{\Delta_{\text{HII}}} d\Delta \, \Delta^2 \, P(\Delta) \equiv \alpha(T)n_e^{\text{II}} R(\Delta_{\text{HII}}) \tag{2.33}$$

The evolution for the mass ionized fraction is then

$$\frac{dF^M(\Delta_{\text{HII}})}{dt} = \frac{\dot{n}_{\text{ph}}}{n_{\text{HII}}^{\text{II}}} - R(\Delta_{\text{HII}})\alpha(T)n_e^{\text{II}} \tag{2.34}$$

The evolution equation essentially tracks the evolution of Δ_{HII} which rises as $F^M(\Delta_{\text{HII}})$ increases with time (i.e., more and more high density regions are getting ionized). Since the mean free path is determined by the high density regions, one should be able to relate it to the value of Δ_{HII} [43]. It is clear that a photon will be able to travel through the low density ionized volume

$$F_V(\Delta_{\text{HII}}) = \int_0^{\Delta_{\text{HII}}} d\Delta \, P(\Delta) \tag{2.35}$$

before being absorbed. When a very high fraction of volume is ionized, one can assume that the fraction of volume filled up by the high density regions is $1 - F_V$, hence their size is proportional to $(1 - F_V)^{1/3}$, and the separation between them along a random line of sight will be proportional to $(1 - F_V)^{-2/3}$, which, in turn, will determine the mean free path. Then one has

$$\lambda_\nu(a) = \frac{\lambda_0}{[1 - F_V(\Delta_{\text{HII}})]^{2/3}} \tag{2.36}$$

where we can fix λ_0 by comparing with low redshift observations like the distribution of Lyman-limit systems [equation (2.23)].

The situation is slightly more complicated when the ionized regions are in the pre-overlap stage. At this stage, a volume fraction $1 - Q_{\text{HII}}$ of the universe is completely neutral (irrespective of the density), while the remaining Q_{HII} fraction of the volume is occupied by ionized regions. However, within this ionized volume, the high density regions (with $\Delta > \Delta_{\text{HII}}$) will still be neutral. Once Q_{HII} becomes unity, all regions with $\Delta < \Delta_{\text{HII}}$ are ionized and the rest are neutral; this can be thought of as the end of reionization. The generalization of equation (2.34), appropriate for this description is given by [43]

$$\frac{d[Q_{\text{HII}} F_M(\Delta_{\text{HII}})]}{dt} = \frac{\dot{n}_{\text{ph}}(z)}{n_{\text{HII}}^{\text{II}}} - Q_{\text{HII}} \alpha_R(T) n_e R(\Delta_{\text{HII}}) \qquad (2.37)$$

Note that there are two unknowns Q_{HII} and $F_M(\Delta_{\text{HII}})$ in equation (2.37) which is impossible to solve without more assumptions. One assumption which is usually made is that Δ_{HII} does not evolve significantly with time in the pre-overlap stage, i.e., it is equal to a critical value Δ_c. This critical density is determined from the the mean separation of the ionizing sources. To have some idea about the value of Δ_c, two arguments have been put forward in the literature: In the first, it is argued that Δ_c is determined by the distribution of sources [43]. When the sources are very numerous, every low-density region (void) can be ionized by sources located at the edges, and hence the overlap of ionized regions can occur (i.e., Q_{HII} approaches unity) when $\Delta_c \sim 1$ is the characteristic overdensity of the thin walls separating the voids. For rare and luminous sources, the mean separation is much larger and hence the value of Δ_c has to be higher before Q_{HII} can be close to unity. In the second approach, it is assumed that the mean free path is determined by the distance between collapsed objects (which manifest themselves as Lyman-limit systems) and hence Δ_c should be similar to the typical overdensities near the boundaries of the collapsed haloes [10, 13]. It usually turns out to be $\sim 50 - 60$ depending on the density profile of the halo. Interestingly, results do not vary considerably as Δ_c is varied from ~ 10 to ~ 100. Once Δ_c is fixed, one can follow the evolution of Q_{HII} until it becomes unity. Following that, we enter the post-overlap stage, where the situation is well-described by equation (2.34).

Of course, the above description is also not fully adequate as there will be a dependence on how far the high density region is from an ionizing source. A dense region which is very close to an ionizing source will be ionized quite early compared to, say, a low-density region which is far away from luminous sources. However, it has been found that the above description gives a reasonable analytical description of the reionization process, particularly for the post-reionization phase. The main advantages in this approach are (i) it takes into account both the pre-overlap and post-overlap phases under a single formalism, (ii) once we have some form for the IGM density distribution $P(\Delta)$, we can calculate the clumping factor and the effective recombination rate self-consistently without introducing any extra parameter; in addition we can also compute the mean free path using one single parameter (λ_0, which can be fixed by comparing with low-redshift observations).

2.3   Modelling of reionization

Given the formalism we have outlined in the previous section, we can now go forward and discuss some other details involved in modelling reionization.

2.3.1   *Reionization sources*

The main uncertainty in any reionization model is to identify the sources. The most natural sources which have been observed to produce ionizing photons are the star-forming galaxies and quasars. Among these, the quasar population is seen to decrease rapidly at $z > 3$ and there is still no evidence of a significant population at higher redshifts. Hence, the most common sources studied in this area are the galaxies.

The subject area of formation of galaxies is quite involved in itself dealing with formation of non-linear structures (haloes and filaments), gas cooling and generation of radiation from stars.

2.3.1.1   *Mass function of collapsed haloes*

The crucial ingredient for galaxy formation is the collapse and virialization of dark matter haloes. This can be adequately described by the Press-Schechter formalism for most purposes. It can be shown that the number density of collapsed objects per unit comoving volume (which is physical volume divided by a^3) within a mass range $(M, M + \mathrm{d}M)$ at an epoch t is given by [50]

$$\frac{\partial n(M,t)}{\partial M}\mathrm{d}M = -\sqrt{\frac{2}{\pi}}\mathrm{e}^{-\nu^2/2}\frac{\bar{\rho}_m}{M}\frac{\mathrm{d}\ln\sigma(M)}{\mathrm{d}\ln M}\frac{\nu}{M}\mathrm{d}M, \qquad (2.38)$$

where $\bar{\rho}_m$ is the comoving density of dark matter, $\sigma(M)$ is defined as the rms mass fluctuation at a mass scale M at $z = 0$, $\nu \equiv \delta_c/[D(t)\sigma(M)]$, $D(t)$ is the growth factor for linear dark matter perturbations and δ_c is the critical overdensity for collapse, usually taken to be equal to 1.69 for a matter-dominated flat universe ($\Omega_m = 1$). This formalism can be extended to calculate the comoving number density of collapsed objects having mass in the range $(M, M + dM)$, which are formed within the time interval $(t_{\text{form}}, t_{\text{form}} + dt'_{\text{form}})$ and observed at a later time t [56, 11] is given by

$$\frac{\partial^2 n(M, t; t_{\text{form}})}{\partial M \partial t_{\text{form}}} dM dt_{\text{form}} = \left.\frac{\partial^2 n(M, t_{\text{form}})}{\partial M \partial t_{\text{form}}}\right|_{\text{form}} \\ \times p_{\text{surv}}[t|t_{\text{form}}] \, dM dt_{\text{form}}$$

(2.39)

where $\partial^2 n(M, t_{\text{form}})/\partial M \partial t_{\text{form}}|_{\text{form}}$ is the formation rate of haloes at t_{form} and $p_{\text{surv}}[t|t_{\text{form}}]$ is their survival probability till time t. Assuming that haloes are destroyed only when they merge to a halo of higher mass, both these quantities can be calculated from the merger rates of haloes. The merger rates can be calculated using detailed properties of Gaussian random field. The quantities can also be calculated in a more simplistic manner by assuming that the merger probability is scale invariant; in that case [56]

$$\left.\frac{\partial^2 n(M, t)}{\partial M \partial t}\right|_{\text{form}} = \frac{\partial^2 n(M, t)}{\partial M \partial t} + \frac{\partial n(M, t)}{\partial M}\frac{\dot{D}}{D}$$
$$= \frac{\partial n(M, t)}{\partial M} \nu^2 \frac{\dot{D}}{D} \qquad (2.40)$$

and

$$p_{\text{surv}}[t|t_{\text{form}}] = \frac{D(t)}{D(t_{\text{form}})}. \qquad (2.41)$$

2.3.1.2 *Star formation rate*

If these dark matter haloes are massive enough to form huge potential wells, the baryonic gas will simply fall into those wells. As the gas begins to settle into the dark matter haloes, mergers will heat it up to the virial temperature via shocks. However, to form galaxies, the gas has to dissipate its thermal energy and cool. If the gas contains only atomic hydrogen, it is unable to cool at temperatures lower than 10^4 K because the atomic hydrogen recombines and cannot be ionized by collisions. The gas can cool effectively for much lower temperatures in the presence of molecules – however, it is not straightforward to estimate the amount of molecules present in the gas at high redshifts. Hence the lower mass cutoff for the haloes which can host star formation will be decided by the cooling efficiency of the baryons.

Let $\dot{M}_*(M, t, t_{\text{form}})$ denote the rate of star formation at time t within a halo of mass M which has formed at t_{form}. Then we can write the cosmic SFR per unit volume at a time t,

$$\dot{\rho}_*(t) = \frac{1}{a^3(t)} \int_0^t dt_{\text{form}} \int_{M_{\text{min}}(t)}^\infty dM' \dot{M}_*(M', t, t_{\text{form}})$$
$$\times \frac{\partial^2 n(M', t; t_{\text{form}})}{\partial M' \partial t_{\text{form}}} \tag{2.42}$$

where the a^{-3} is included to covert from comoving to physical volume. The lower mass cutoff $M_{\text{min}}(t)$ at a given epoch is decided by the cooling criteria as explained above. However, once reionization starts and regions are re-heated by photoheating, the value of $M_{\text{min}}(t)$ is set by the photoionization temperature $\simeq 10^4$ K. This can further suppress star formation in low mass haloes and is known as radiative feedback. We shall discuss this later in this section.

The form of $\dot{M}_*(M, t, t_{\text{form}})$ contains information about various cooling and star-forming processes and hence is quite complex to deal with. It can be obtained from semi-analytical modelling of galaxy formation [60] or constrained from observations of galaxy luminosity function [54]. A very simple assumption that is usually made for modelling reionization is that the duration of star formation is much less than the Hubble time $H^{-1}(t)$ which is motivated by the fact that most of the ionizing radiation is produced by hot stars which have shorter lifetime. In that case, $\dot{M}_*(M, t, t_{\text{form}})$ can be approximated as

$$\dot{M}_*(M, t, t_{\text{form}}) \approx M_* \delta_D(t - t_{\text{form}}) = f_* \frac{\bar{\rho}_b}{\bar{\rho}_m} M \delta_D(t - t_{\text{form}}) \tag{2.43}$$

where $\bar{\rho}_b / \bar{\rho}_m M$ is the mass of baryons within the halo and f_* is the fraction of baryonic mass which has been converted into stars. The cosmic SFR per comoving volume is then

$$\dot{\rho}_*(t) = \frac{1}{a^3(t)} \frac{\bar{\rho}_b}{\bar{\rho}_m} \int_{M_{\text{min}}(t)}^\infty dM' f_* M' \left. \frac{\partial^2 n(M', t)}{\partial M' \partial t} \right|_{\text{form}} \tag{2.44}$$

One should keep in mind that many details of the star-formation process have been encoded within a single parameter f_*. This should, in principle, be a function of both halo mass M and time t. However, it is not clear at all what the exact dependencies should be. Given such uncertainties, it is usual to take it as a constant.

In addition, one finds that the merger rate of haloes at high redshifts is much less than the formation rate (which follows if $\nu \gg 1$), hence $\partial^2 n(M,t)/\partial M \partial t|_{\text{form}} \approx \partial^2 n(M,t)/\partial M \partial t$. Then, one can write the SFR in terms of the fraction of collapsed mass in haloes more massive than $M_{\text{min}}(t)$

$$f_{\text{coll}}(t) = \frac{1}{\bar{\rho}_m} \int_{M_{\text{min}}(t)}^{\infty} \mathrm{d}M' \, M' \frac{\partial n(M',t)}{\partial M'}$$

$$= \operatorname{erfc}\left[\frac{\delta_c}{\sqrt{2} \, D(t)\sigma(M_{\text{min}})}\right] \tag{2.45}$$

as

$$\dot{\rho}_*(t) = f_* \frac{\bar{\rho}_b}{a^3(t)} \frac{\mathrm{d}f_{\text{coll}}(t)}{\mathrm{d}t} \tag{2.46}$$

2.3.1.3   *Production of ionizing photons*

Given the SFR, we can calculate the emissivity of galaxies, or equivalently the rate of ionizing photons in the IGM per unit volume per unit frequency range:

$$\dot{n}_{\nu,\text{ph}}(t) = f_{\text{esc}} \left[\frac{\mathrm{d}N_\nu}{\mathrm{d}M_*}\right] \dot{\rho}_*(t) \tag{2.47}$$

where $\mathrm{d}N_\nu/\mathrm{d}M_*$ gives the number of photons emitted per frequency range per unit mass of stars and f_{esc} is the fraction of ionizing photons which escape from the star forming haloes into the IGM. The emissivity is simply $\epsilon_\nu = h_p \nu \dot{n}_{\nu,\text{ph}}$.

Given the spectra of stars of different masses in a galaxy, and their Initial Mass Function (IMF), $\mathrm{d}N_\nu/\mathrm{d}M_*$ can be computed in a straightforward way using "population synthesis" codes [35, 7]. The IMF and spectra will depend on the details of star formation (burst formation or continuous) and metallicity. In fact, it is possible that there are more than one population of stellar sources which have different $\mathrm{d}N_\nu/\mathrm{d}M_*$. For example, there are strong indications, both from numerical simulations and analytical arguments [6], that the first generation stars were metal-free, and hence massive, with a very different kind of IMF and spectra than the stars we observe today [57]; they are known as the Population III (PopIII) stars.

A fraction of photons produced in a galaxy would be consumed in ionizing the neutral matter within the galaxy itself. Hence only a fraction of photons escape into the IGM and is available for reionization, which is encoded in the parameter f_{esc}. This parameter is again not very well modelled and its observed value is also quite uncertain; typical values assumed are ~ 0.1. It is most likely f_{esc}, like f_*, is also a function of halo mass and the time of halo formation, however since the dependences are not well understood, it is taken to be a constant.

The total number of ionizing photons is then obtained by integrating the above quantity over all energies above the ionization threshold:

$$\dot{n}_{ph}(t) = \int_{\nu_{HI}}^{\infty} d\nu \; \dot{n}_{\nu,ph}(t) = N_{ion} n_b \frac{d f_{coll}(t)}{dt} \qquad (2.48)$$

where n_b is the total baryonic number density in the IGM (equal to n_H if we ignore the presence of helium) and

$$N_{ion} \equiv f_* f_{esc} m_p \int_{\nu_{HI}}^{\infty} d\nu \left[\frac{dN_\nu}{dM_*} \right] \qquad (2.49)$$

is the number of photons entering the IGM per baryon in collapsed objects. In case there are more than one population of stars, one has to use different values of N_{ion} for the different populations.

2.3.1.4 *Feedback processes*

The moment there is formation of stars and other luminous bodies, they start to affect the subsequent formation of structures – this is known as feedback [17]. The process is intrinsically non-linear and hence quite complex to model. The feedback processes can be categorized roughly into three categories.

The first of them is the radiative feedback which is associated with the radiation from first stars which can heat up the medium and can photoionize atoms and/or photodissociate molecules. Once the first galaxies form stars, their radiation will ionize and heat the surrounding medium, increasing the mass scale (often referred to as the *filtering mass* [27]) above which baryons can collapse into haloes within those regions. The minimum mass of haloes which are able to cool is thus much higher in ionized regions than in neutral ones. Since the IGM is multi-phase in the pre-overlap phase, one needs to take into account the heating of the ionized regions right from the beginning of reionization. In principle, this can be done self-consistently from the evolution of the temperature of the ionized region in equation (2.3).

The low mass haloes can be subjected to mechanical feedback too, which is mostly due to energy injection via supernova explosion and winds. This can expel the gas from the halo and suppress star formation. As in the case for radiative feedback, one can parametrize this through the minimum mass parameter $M_{min}(t)$.

Finally, we also have chemical feedback where the stars expel metals into the medium and hence change the chemical composition. This would mean that the subsequent formation of stars could be in a completely different environment and hence the nature of stars would be highly different.

2.3.1.5 *Quasars*

Besides the stellar sources, a population of accreting black holes or quasars are also known to produce significant amount of ionizing radiation. Hence it is possible that they too have contributed to reionization. The fraction of their contribution would depend on the number of quasars produced at a particular redshift. Observationally, the luminosity function of quasars is quite well-probed till a redshift $z \sim 6$ [53]. It turns out that the population peaks around $z \approx 3$ and decreases for higher redshifts. Hence their contribution at higher redshifts is highly debated.

The difference between reionization by stellar sources and quasars lie in the fact that quasars produce significant number of high energy photons compared to stars. This would imply that quasars can contribute significantly to double-reionization of helium (which requires photons with energies > 54.4 eV, not seen in galaxies). In addition, quasars produce significant amount of X-ray radiation. Since the absorption cross section of neutral hydrogen varies with frequency approximately as ν^{-3}, the mean free path for photons with high energies would be very large. A simple calculation will show that for photons with energies above 100–200 eV, the mean free path would be larger than the typical separation between collapsed structures [39] (the details would depend upon the redshift and exact description of collapsed haloes). These photons would not be associated with any particular source at the moment when they are absorbed, and thus would ionize the IGM in a more homogeneous manner (as opposed to the overlapping bubble picture for UV sources).

2.3.2 *Illustration of a semi-analytical model*

The physics described above in the preceding sections can all be combined to construct a semi-analytical model for studying the thermal and ionization history of the IGM. We shall give an explicit example of one such model [13, 15] whose main features are: The model accounts for IGM inhomogeneities by adopting a lognormal distribution for $P(\Delta)$; reionization is said to be complete once all the low-density regions (say, with overdensities $\Delta < \Delta_c \sim$

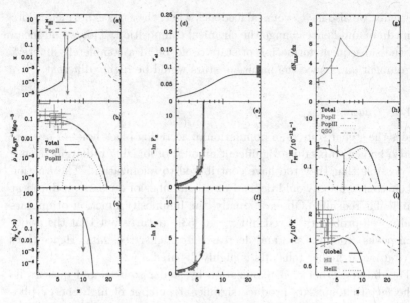

Fig. 2.2 Comparison of analytical model predictions with observations for the best-fit model. The different panels indicate: (a) The volume-averaged neutral hydrogen fraction $x_{\rm HI}$, with observational lower limit from QSO absorption lines at $z = 6$ and upper limit from Lyα emitters at $z = 6.5$ (shown with arrows). In addition, the ionized fraction x_e is shown by the dashed line. (b) SFR $\dot{\rho}_*$ for different stellar populations. (c) The number of source counts above a given redshift, with the observational upper limit from NICMOS HUDF is shown by the arrow. The contribution to the source count is zero at low redshifts because of the J-dropout selection criterion. (d) Electron scattering optical depth, with observational constraint from WMAP 3-year data release. (e) Lyα effective optical depth. (f) Lyβ effective optical depth. (g) Evolution of Lyman-limit systems. (h) Photoionization rates for hydrogen, with estimates from numerical simulations (shown by points with error-bars; [4]). (i) Temperature of the mean density IGM.

60) are ionized. The ionization and thermal histories of neutral, HII and HeIII regions are followed simultaneously and self-consistently, treating the IGM as a multi-phase medium. Three types of reionization sources have been assumed: (i) metal-free (i.e. PopIII) stars having a Salpeter IMF in the mass range $1 - 100 M_\odot$: they dominate the photoionization rate at high redshifts; (ii) PopII stars with sub-solar metallicities also having a Salpeter IMF in the mass range $1-100 M_\odot$; (iii) quasars, which are significant sources of hard photons at $z \lesssim 6$; they have negligible effects on the IGM at higher redshifts.

As discussed earlier, reionization is accompanied by various feedback processes, which can affect subsequent star formation. In this model, radiative feedback is computed self-consistently from the evolution of the thermal properties of the IGM. Furthermore, the chemical feedback inducing the PopIII $\rightarrow$ PopII transition is implemented using a merger-tree "genetic" approach which determines the termination of PopIII star formation in a metal-enriched halo [59].

The predictions of the model are compared with a wide range of observational data sets, namely, (i) redshift evolution of Lyman-limit absorption systems dN_{LL}/dz [66], (ii) the effective optical depths $\tau_{eff} \equiv -\ln F$ for Lyα and Lyβ absorption in the IGM [61], (iii) electron scattering optical depth $\tau_{el} = \sigma_T c \int dt\, n_e$ (where σ_T is the Thomson scattering cross section) as measured from CMBR experiments [63], (iv) temperature of the mean intergalactic gas [58], (v) cosmic star formation history $\dot{\rho}_*$ [44], and (vi) source number counts at $z \approx 10$ from NICMOS HUDF [5].

The data constrain the reionization scenario quite tightly. We find that hydrogen reionization starts at $z \approx 15$ driven by metal-free (PopIII) stars, and it is 90 per cent complete by $z \approx 8$. After a rapid initial phase, the growth of the volume filled by ionized regions slows down at $z \lesssim 10$ due to the combined action of chemical and radiative feedback, making reionization a considerably extended process completing only at $z \approx 6$. The number of photons per hydrogen at the end of reionization at $z \approx 6$ is only a few, which implies that reionization occurred in a "photon-starved" manner [3].

2.4 Current status and future

In this section, let us review the current status of various approaches to studying reionization and their future prospects.

2.4.1 *Simulations*

Though the analytical studies mentioned above allow us to develop a good understanding of the different processes involved in reionization, they can take into account the physical processes only in some approximate sense. In fact, a detailed and complete description of reionization would require locating the ionizing sources, resolving the inhomogeneities in the IGM, following the scattering processes through detailed radiative transfer, and so on. Numerical simulations, in spite of their limitations, have been of immense importance in these areas.

The ionizing photons during early stages of reionization mostly originate from smaller haloes which are far more numerous than the larger galaxies at high redshifts. The need to resolve such small structures requires the simulation boxes to have high enough resolution. On the other hand, these ionizing sources were strongly clustered at high redshifts and, as a consequence, the ionized regions they created are expected to overlap and grow to very large sizes, reaching upto tens of Mpc. As already discussed, the many orders of magnitude difference between these length scales demand extremely high computing power from any simulations designed to study early structure formation from the point of view of reionization.

To simulate reionization, one usually runs a N-body simulation (either dark matter only or including baryons) to generate the large-scale density field, identifies haloes within the density field and assign ionizing photons to the haloes using a assumption like equation (2.48). It turns out that the most difficult step is to solve the radiative transfer equation and study the growth of ionized regions. In principle, one could solve equation (2.4) directly for the intensity at every point in the seven-dimensional $(t, \mathbf{x}, \mathbf{n}, \nu)$ space, given the absorption coefficient and the emissivity. However, the high dimensionality of the problem makes the solution of the complete radiative transfer equation well beyond our capabilities, particularly since we do not have any obvious symmetries in the problem and often need high spatial and angular resolution in cosmological simulations. At present, most of the simulations do not have enough resolution to reliably identify the low mass $\sim 10^8 M_\odot$ sources which were probably responsible for early stages of reionization. Also, there are difficulties in resolving the small scale structures which contribute significantly to the clumpiness in the IGM and hence extend the reionization process.

2.4.2 *Various observational probes*

Finally, we review certain observations which shape our understanding of reionization.

2.4.2.1 *Absorption spectra of high redshift sources*

We have already discussed that the primary evidence for reionization comes from absorption spectra of quasars (Lyα forest) at $z < 6$. We have also discussed that the effective optical depth of Lyα photons becomes significantly large at $z \gtrsim 6$ implying regions with high transmission in the Lyα forest becoming rare at high redshifts [20, 68]. Therefore the standard methods of analyzing the Lyα forest (like the probability distribution function and power spectrum) are not very effective. Amongst alternate methods, one can use the the distribution of dark gaps [18, 62] which are defined as contiguous regions of the spectrum having an optical depth above a threshold value [62, 20]. It has been found that the current observations constrain the neutral hydrogen fraction $x_{HI} < 0.36$ at $z = 6.3$ [25]. It is expected that the SDSS and Palomar-Quest survey [37] would detect ~ 30 QSOs at these redshifts within the next few years and hence we expect robust conclusions from such studies in very near future.

Like quasars, one can also use absorption spectra of other high redshift energetic sources like gamma ray bursts (GRBs) and supernovae. In fact, analyses using the damping wing effects of the Voigt profile have been already performed on the GRB detected at a redshift $z = 6.3$, and the wing shape is well-fit by a neutral fraction $x_{HI} < 0.17$ [67]. The dark gap width distribution gives a similar constraint $x_{HI} = (6.4\pm0.3) \times 10^{-5}$ [26]. In order to obtain more stringent limits on reionization, it is important to increase the sample size of $z > 6$ GRBs.

In addition to hydrogen reionization, the Lyα forest in the quasar absorption lines at $z \approx 3$ can also be used for studying reionization of singly ionized helium to doubly ionized state (the reionization of neutral helium to singly ionized state follows hydrogen for almost all types of sources). The helium reionization coincides with the rise in quasar population at $z \sim 3$ and it effects the thermal history of the IGM at these redshifts. However, there are various aspects of the observation that are not well understood and requires much detailed modelling of helium reionization.

2.4.2.2 *CMBR observations*

As we have discussed already, the first evidence for an early reionization epoch came from the CMBR polarization data. This data is going to be much more precise in future with experiments like PLANCK, and is expected to improve the constraints on τ_{el}. With improved statistical errors, it might be possible to distinguish between different evolutions of the ionized fraction, particularly with E-mode polarization auto-correlation, as is found from theoretical calculations [31, 8]. An alternative option to probe reionization through CMBR is through the small scale observations of temperature anisotropies. It has been well known that the scattering of the CMBR photons by the bulk motion of the electrons in clusters gives rise to a signal at large multipoles $\ell \sim 1000$, known as the kinetic Sunyaev Zel-dovich (SZ) effect. Such a signal can also originate from the fluctuations in the distribution of free electrons arising from cosmic reionization. It turns out that for reionization, the signal is dominated by the patchiness in the n_e-distribution. Now, in most scenarios of reionization, it is expected that the distribution of neutral hydrogen would be quite patchy in the pre-overlap era, with the ionized hydrogen mostly contained within isolated bubbles. The amplitude of this signal is significant around $\ell \sim 1000$ and is usually comparable to or greater than the signal arising from standard kinetic SZ effect. Theoretical estimates of the signal have been performed for various reionization scenarios, and it has been predicted that the experiment can be used for constraining reionization history [55, 42]. Also, it is possible to have an idea about the nature of reionization sources, as the signal from UV sources, X-ray sources and decaying particles are quite different. With multi-frequency experiments like Atacama Cosmology Telescope (ACT)[2] and South Pole Telescope (SPT)[3] coming up in near future, this promises to put strong constraints on the reionization scenarios.

2.4.2.3 *Lyα emitters*

In recent years, a number of groups have studied star-forming galaxies at $z \sim 6 - 7$, and measurements of the Lyα emission line luminosity function evolution provide another useful observational constraint [40, 64]. While the quasar absorption spectra probe the neutral hydrogen fraction regime $x_{HI} \leq 0.01$, this method is sensitive to the range $x_{HI} \sim 0.1 - 1.0$. Lyα emission from galaxies is expected to be suppressed at redshifts beyond reionization because of the absorption due to neutral hydrogen, which clearly affects the evolution of the luminosity function of such Lyα emitters at high redshifts [30, 40, 22]. Thus a comparison of the luminosity functions at different redshifts could be used for constraining the reionization. Through a simple analysis, it was found that the luminosity functions at $z = 5.7$ and $z = 6.5$ are statistically consistent with one another thus implying that reionization was largely complete at $z \approx 6.5$. More sophisticated calculations on the evolution of the luminosity function of Lyα emitters [40, 24, 30] suggest that the neutral fraction of hydrogen at $z = 6.5$ should be less than 50 per cent [41]. Unfortunately, the analysis of the Lyα emitters at high redshifts is complicated by various factors like the velocity of the sources

---

[2]http://www.hep.upenn.edu/act/
[3]http://spt.uchicago.edu/

with respect to the surrounding IGM, the density distribution and the size of ionized regions around the sources and the clustering of sources. It is thus extremely important to have detailed models of Lyα emitting galaxies in order to use them for constraining reionization.

2.4.2.4 *Sources of reionization*

As we discussed earlier, a major challenge in our understanding of reionization depends on our knowledge of the sources, particularly at high redshifts. As we understand at present, neither the bright $z > 6$ quasars discovered by the SDSS group [21] nor the faint AGN detected in X-ray observations [1] produce enough photons to reionize the IGM. The discovery of star-forming galaxies at $z > 6.5$ [32, 34, 33] has resulted in speculation that early galaxies produce bulk of the ionizing photons for reionization. Unfortunately, there are significant uncertainties in constraining the amount of ionizing radiation at these redshifts because the bulk of ionizing photons could be produced by faint sources which are beyond the present sensitivities. In fact, some models have predicted that the $z > 7$ sources identified in these surveys are relatively massive ($M \approx 10^9 M_\odot$) and rare objects which are only marginally ($\approx 1\%$) contributing to the reionization photon budget [16]. A much better prospect of detecting these sources would be through the Ultra-Deep Imaging Survey using the future telescope JWST.

2.4.2.5 *21cm observations*

Perhaps the most promising prospect of detecting the fluctuations in the neutral hydrogen density during the (pre-)reionization era is through the 21 cm emission experiments [23], some of which are already taking data (GMRT [4], 21CMA [5]), and some are expected in future (MWA [6], LOFAR [7], SKA [8]). The basic principle which is central to these experiments is the neutral hydrogen hyperfine transition line at a rest wavelength of 21 cm. This line, when redshifted, is observable in radio frequencies (~ 150 MHz for $z \sim 10$) as a brightness temperature:

$$\delta T_b(z, \hat{n}) = \frac{T_S - T_{CMB}}{1 + z} \frac{3c^3 \bar{A}_{10} n_{HI}(z, \hat{n})(1 + z)^3}{16 k_{boltz} \nu_0^2 T_S H(z)} \qquad (2.50)$$

---

[4]http://www.gmrt.ncra.tifr.res.in
[5]http://web.phys.cmu.edu/~past/
[6]http://www.haystack.mit.edu/arrays/MWA
[7]http://www.lofar.org
[8] http://www.skatelescope.org/

where T_S is the spin temperature of the gas, $T_{CMB} = 2.76(1 + z)$ K is the CMBR temperature, A_{10} is the Einstein coefficient and $\nu_0 = 1420$ MHz is the rest frequency of the hyperfine line.

The observability of this brightness temperature against the CMB background will depend on the relative values of T_S and T_{CMB}. Depending on which processes dominate at different epochs, T_S will couple either to radiation (T_{CMB}) or to matter (determined by the kinetic temperature T_k) [51]. Almost in all models of reionization, the most interesting phase for observing the 21 cm radiation is $6 \lesssim z \lesssim 20$. This is the phase where the IGM is suitably heated to temperatures much higher than CMB (mostly due to X-ray heating [9]) thus making it observable in emission. In that case, we have $\delta T_b \propto n_{HI}/n_H$, which means that the observations would directly probe the neutral hydrogen density in the Universe. Furthermore, this is the era when the bubble-overlapping phase is most active, and there is substantial neutral hydrogen to generate a strong enough signal. At low redshifts, after the IGM is reionized, n_{HI} falls by orders of magnitude and the 21 cm signal vanishes except in the high density neutral regions. Since the observations directly probe the neutral hydrogen density, one can use it to probe the detailed topology of the ionized regions in the pre-overlap phase. It is therefore essential to model the clustering of the sources accurately so as to predict the reionization topology.

There are essentially two complementary approaches to studying reionization using 21 cm signal. The first one is through global statistical properties of the neutral hydrogen signal, like the power spectrum. The second one is to directly detect the ionized bubbles around sources, either through blind surveys or via targetted observations.

The major difficulty in obtaining the cosmological signal from these experiments is that it is expected to be only a small contribution buried deep in the emission from other astrophysical sources (foregrounds) and in the system noise. It is thus a big challenge to detect the signal which is of cosmological importance from the other contributions that are orders of magnitude larger. Once such challenges are dealt with, this probe will be the strongest probe for not only reionization, but of the matter distribution at very small scales during the dark ages.

2.5 Concluding remarks

We have discussed the analytical approaches to model different aspects of reionization which will help in understanding the most relevant physical processes. In an explicit example, we have shown how to apply this formalism for constraining the reionization history using a variety of observational data. These constraints imply that reionization is an extended process over a redshift range $15 > z > 6$. It is most likely driven by the first sources which form in small mass haloes. However, there are still uncertainties about the exact nature of these sources and the detailed topology of ionized regions. Such details are going to be addressed in near future as new observations, both space-borne and ground-based, are likely to settle these long-standing questions. From the theoretical point of view, it is thereby important to develop detailed analytical and numerical models to extract the maximum information about the physical processes relevant for reionization out of the expected large and complex data sets.

References

[1] Barger, A. J., Cowie, L. L., Capak, P., Alexander, D. M., Bauer, F. E., Brandt, W. N., Garmire, G. P. and Hornschemeier, A. E. (2003). Very High Redshift X-Ray-selected Active Galactic Nuclei in the Chandra Deep Field-North, *ApJ* **584**, pp. L61–L64.

[2] Barkana, R. and Loeb, A. (2001). In the beginning: the first sources of light and the reionization of the universe, *Phys. Rep.* **349**, pp. 125–238, doi:10.1016/S0370-1573(01)00019-9.

[3] Bolton, J. S. and Haehnelt, M. G. (2007). The observed ionization rate of the intergalactic medium and the ionizing emissivity at $z >=$ 5: evidence for a photon-starved and extended epoch of reionization, *MNRAS* **382**, pp. 325–341, doi:10.1111/j.1365-2966.2007.12372.x.

[4] Bolton, J. S., Haehnelt, M. G., Viel, M. and Springel, V. (2005). The Lyman α forest opacity and the metagalactic hydrogen ionization rate at $z \sim$ 2-4, *MNRAS* **357**, pp. 1178–1188, doi:10.1111/j.1365-2966.2005.08704.x.

[5] Bouwens, R. J., Illingworth, G. D., Thompson, R. I. and Franx, M. (2005). Constraints on $z \sim 10$ Galaxies from the Deepest Hubble Space Telescope NICMOS Fields, *ApJ* **624**, pp. L5–L8.

[6] Bromm, V., Kudritzki, R. P. and Loeb, A. (2001). Generic Spectrum

and Ionization Efficiency of a Heavy Initial Mass Function for the First Stars, *ApJ* **552**, pp. 464–472, doi:10.1086/320549.

[7] Bruzual, G. and Charlot, S. (2003). Stellar population synthesis at the resolution of 2003, *MNRAS* **344**, pp. 1000–1028.

[8] Burigana, C., Popa, L. A., Salvaterra, R., Schneider, R., Choudhury, T. R. and Ferrara, A. (2008). Cosmic microwave background polarization constraints on radiative feedback, *MNRAS* **385**, pp. 404–410, doi:10.1111/j.1365-2966.2008.12845.x.

[9] Chen, X. and Miralda-Escudé, J. (2004). The Spin-Kinetic Temperature Coupling and the Heating Rate due to Lyα Scattering before Reionization: Predictions for 21 Centimeter Emission and Absorption, *ApJ* **602**, pp. 1–11.

[10] Chiu, W. A., Fan, X. and Ostriker, J. P. (2003). Combining Wilkinson Microwave Anisotropy Probe and Sloan Digital Sky Survey Quasar Data on Reionization Constrains Cosmological Parameters and Star Formation Efficiency, *ApJ* **599**, pp. 759–772, doi:10.1086/379318.

[11] Chiu, W. A. and Ostriker, J. P. (2000). A Semianalytic Model for Cosmological Reheating and Reionization Due to the Gravitational Collapse of Structure, *ApJ* **534**, pp. 507–532.

[12] Choudhury, T. R. (2009). Analytical Models of the Intergalactic Medium and Reionization, *Curr. Sci.* **97**, pp. 841–857.

[13] Choudhury, T. R. and Ferrara, A. (2005). Experimental constraints on self-consistent reionization models, *MNRAS* **361**, pp. 577–594.

[14] Choudhury, T. R. and Ferrara, A. (2006). Physics of cosmic reionization, in R. Fabbri (ed.), *Cosmic Polarization* (Research Signpost, India), p. 205.

[15] Choudhury, T. R. and Ferrara, A. (2006). Updating reionization scenarios after recent data, *MNRAS* **371**, pp. L55–L59.

[16] Choudhury, T. R. and Ferrara, A. (2007). Searching for the reionization sources, *MNRAS* **380**, pp. L6–L10.

[17] Ciardi, B. and Ferrara, A. (2005). The First Cosmic Structures and Their Effects, *Space Sci. Rev.* **116**, pp. 625–705.

[18] Croft, R. A. C. (1998). Characterization of Lyman Alpha Spectra and Predictions of Structure Formation Models: A Flux Statistics Approach, in *Olinto A. V., Frieman J. A., Schramm D. N. ed., Eighteenth Texas Symposium on Relativistic Astrophysics.* (World Scientific, River Edge, N. J.), pp. 664–+.

[19] Fan, X., Carilli, C. L. and Keating, B. (2006). Observational Constraints on Cosmic Reionization, *ARA&A* **44**, pp. 415–462.

[20] Fan, X., Strauss, M. A., Becker, R. H., White, R. L., Gunn, J. E., Knapp, G. R., Richards, G. T., Schneider, D. P., Brinkmann, J. and Fukugita, M. (2006). Constraining the Evolution of the Ionizing Background and the Epoch of Reionization with z~6 Quasars. II. A Sample of 19 Quasars, *AJ* **132**, pp. 117–136.

[21] Fan, X., Strauss, M. A., Richards, G. T., Hennawi, J. F., Becker, R. H., White, R. L., Diamond-Stanic, A. M., onley, J. L. D., Jiang, L., Kim, J. S., Vestergaard, M., Young, J. E., Gunn, J. E., Lupton, R. H., Knapp, G. R., Schneider, D. P., Brandt, W. N., Bahcall, N. A., Barentine, J. C., Brinkmann, J., Brewington, H. J., ukugita, M. F., Harvanek, M., Kleinman, S. J., Krzesinski, J., Long, D., eilsen, E. H. N., Jr., Nitta, A., Snedden, S. A. and Voges, W. (2005). Preprint: astro-ph/0512080.

[22] Furlanetto, S. R., Hernquist, L. and Zaldarriaga, M. (2004). Constraining the topology of reionization through Lyα absorption, *MNRAS* **354**, pp. 695–707.

[23] Furlanetto, S. R., Oh, S. P. and Briggs, F. H. (2006). Cosmology at low frequencies: The 21 cm transition and the high-redshift Universe, *Phys. Rep.* **433**, pp. 181–301.

[24] Furlanetto, S. R., Zaldarriaga, M. and Hernquist, L. (2006). The effects of reionization on Lyα galaxy surveys, *MNRAS* **365**, pp. 1012–1020.

[25] Gallerani, S., Ferrara, A., Fan, X. and Choudhury, T. R. (2008). Glimpsing through the high-redshift neutral hydrogen fog, *MNRAS* **386**, pp. 359–369, doi:10.1111/j.1365-2966.2008.13029.x.

[26] Gallerani, S., Salvaterra, R., Ferrara, A. and Choudhury, T. R. (2008). Testing reionization with gamma-ray burst absorption spectra, *MNRAS* **388**, pp. L84–L88, doi:10.1111/j.1745-3933.2008.00504.x.

[27] Gnedin, N. Y. and Hui, L. (1998). Probing the Universe with the Lyalpha forest - I. Hydrodynamics of the low-density intergalactic medium, *MNRAS* **296**, pp. 44–55.

[28] Gunn, J. E. and Peterson, B. A. (1965). On the Density of Neutral Hydrogen in Intergalactic Space. *ApJ* **142**, pp. 1633–1641.

[29] Haardt, F. and Madau, P. (1996). Radiative Transfer in a Clumpy Universe. II. The Ultraviolet Extragalactic Background, *ApJ* **461**, p. 20.

[30] Haiman, Z. and Cen, R. (2005). Constraining Reionization with the Evolution of the Luminosity Function of Lyα Emitting Galaxies, *ApJ* **623**, pp. 627–631.

[31] Holder, G. P., Haiman, Z., Kaplinghat, M. and Knox, L. (2003). The Reionization History at High Redshifts. II. Estimating the Optical

Depth to Thomson Scattering from Cosmic Microwave Background Polarization, *ApJ* **595**, pp. 13–18.

[32] Hu, E. M., Cowie, L. L., McMahon, R. G., Capak, P., Iwamuro, F., Kneib, J.-P., Maihara, T. and Motohara, K. (2002). A Redshift z=6.56 Galaxy behind the Cluster Abell 370, *ApJ* **568**, pp. L75–L79.

[33] Kneib, J.-P., Ellis, R. S., Santos, M. R. and Richard, J. (2004). A Probable z~7 Galaxy Strongly Lensed by the Rich Cluster A2218: Exploring the Dark Ages, *ApJ* **607**, pp. 697–703.

[34] Kodaira, K., Taniguchi, Y., Kashikawa, N., Kaifu, N., Ando, H., Karoji, H., Ajiki, M., Akiyama, M., Aoki, K., Doi, M., Fujita, S. S., Furusawa, H., Hayashino, T., Imanishi, M., Iwamuro, F., Iye, M., Kawabata, K. S., Kobayashi, N., Kodama, T., Komiyama, Y., Kosugi, G., Matsuda, Y., Miyazaki, S., Mizumoto, Y., Motohara, K., Murayama, T., Nagao, T., Nariai, K., Ohta, K., Ohyama, Y., Okamura, S., Ouchi, M., Sasaki, T., Sekiguchi, K., Shimasaku, K., Shioya, Y., Takata, T., Tamura, H., Terada, H., Umemura, M., Usuda, T., Yagi, M., Yamada, T., Yasuda, N. and Yoshida, M. (2003). The Discovery of Two Lyman α Emitters beyond Redshift 6 in the Subaru Deep Field, *PASJ* **55**, pp. L17–L21.

[35] Leitherer, C., Schaerer, D., Goldader, J. D., Delgado, R. M. G., Robert, C., Kune, D. F., de Mello, D. F., Devost, D. and Heckman, T. M. (1999). Starburst99: Synthesis Models for Galaxies with Active Star Formation, *ApJS* **123**, pp. 3–40.

[36] Loeb, A. and Barkana, R. (2001). The Reionization of the Universe by the First Stars and Quasars, *ARA&A* **39**, p. 19.

[37] López-Cruz, O., Djorgovski, S. G., Carrasco, L., Baltay, C., Recillas, E., Mayya, Y. D., Escobedo, G. A., Castillo, E., Mahabal, A., Thompson, D. J., Rabinowitz, D., Bauer, A., Graham, M., Williams, R., Rengstorf, A., Brunner, R., Andrews, P., Ellman, N., Duffau, S., Lauer, R., Bogosavljevic, M., Musser, J., Mufson, S. and Gebhard, M. (2005). A Search for z ~ 6 QSOs in the Palomar-Quest Sky Survey, in A. M. Hidalgo-Gámez, J. J. González, J. M. Rodríguez Espinosa and S. Torres-Peimbert (eds.), *Revista Mexicana de Astronomia y Astrofisica Conference Series* (Instituto de Astronoma, Universidad Nacional Autnoma de Mxico, Mexico), pp. 164–169.

[38] Madau, P., Haardt, F. and Rees, M. J. (1999). Radiative Transfer in a Clumpy Universe. III. The Nature of Cosmological Ionizing Sources, *ApJ* **514**, p. 648.

[39] Madau, P., Rees, M. J., Volonteri, M., Haardt, F. and Oh, S. P. (2004).

Early Reionization by Miniquasars, *ApJ* **604**, pp. 484–494.

[40] Malhotra, S. and Rhoads, J. E. (2004). Luminosity Functions of Lyα Emitters at Redshifts z=6.5 and z=5.7: Evidence against Reionization at z<=6.5, *ApJ* **617**, pp. L5–L8.

[41] Malhotra, S. and Rhoads, J. E. (2006). The Volume Fraction of Ionized Intergalactic Gas at Redshift z=6.5, *ApJ* **647**, pp. L95–L98, doi:10. 1086/506983.

[42] McQuinn, M., Furlanetto, S. R., Hernquist, L., Zahn, O. and Zaldarriaga, M. (2005). The Kinetic Sunyaev-Zel'dovich Effect from Reionization, *ApJ* **630**, pp. 643–656.

[43] Miralda-Escudé, J., Haehnelt, M. and Rees, M. J. (2000). Reionization of the Inhomogeneous Universe, *ApJ* **530**, pp. 1–16.

[44] Nagamine, K., Cen, R., Hernquist, L., Ostriker, J. P. and Springel, V. (2004). Is There a Missing Galaxy Problem at High Redshift? *ApJ* **610**, pp. 45–50.

[45] Padmanabhan, T. (2002). *Theoretical Astrophysics, Volume III: Galaxies and Cosmology* (Cambridge, England: Cambridge University Press, UK).

[46] Paresce, F., McKee, C. F. and Bowyer, S. (1980). Galactic and extragalactic contributions to the far-ultraviolet background, *ApJ* **240**, pp. 387–400, doi:10.1086/158244.

[47] Peacock, J. A. (1999). *Cosmological Physics* (Cambridge, UK: Cambridge University Press, UK).

[48] Peebles, P. J. E. (1993). *Principles of physical cosmology* (Princeton, NJ: Princeton University Press, USA).

[49] Petitjean, P., Bergeron, J., Carswell, R. F. and Puget, J. L. (1993). Detailed structure of expanding photoionized Ly-alpha clouds, *MNRAS* **260**, pp. 67–76.

[50] Press, W. H. and Schechter, P. (1974). Formation of galaxies and clusters of galaxies by self-similar gravitational condensation, *ApJ* **187**, pp. 425–438.

[51] Pritchard, J. R. and Loeb, A. (2008). Evolution of the 21cm signal throughout cosmic history, *Phys. Rev. D* **78**, 10, pp. 103511–+, doi: 10.1103/PhysRevD.78.103511.

[52] Rauch, M. (1998). The Lyman Alpha Forest in the Spectra of QSOs, *ARA&A* **36**, p. 267.

[53] Richards, G. T., Strauss, M. A., Fan, X., Hall, P. B., Jester, S., Schneider, D. P., Vanden Berk, D. E., Stoughton, C., Anderson, S. F., Brunner, R. J., Gray, J., Gunn, J. E., Ivezić, Ž., Kirkland, M. K., Knapp,

G. R., Loveday, J., Meiksin, A., Pope, A., Szalay, A. S., Thakar, A. R., Yanny, B., York, D. G., Barentine, J. C., Brewington, H. J., Brinkmann, J., Fukugita, M., Harvanek, M., Kent, S. M., Kleinman, S. J., Krzesiński, J., Long, D. C., Lupton, R. H., Nash, T., Neilsen, E. H., Jr., Nitta, A., Schlegel, D. J. and Snedden, S. A. (2006). The Sloan Digital Sky Survey Quasar Survey: Quasar Luminosity Function from Data Release 3, *AJ* **131**, pp. 2766–2787, doi:10.1086/503559.

[54] Samui, S., Srianand, R. and Subramanian, K. (2007). Probing the star formation history using the redshift evolution of luminosity functions, *MNRAS* **377**, pp. 285–299, doi:10.1111/j.1365-2966.2007.11603.x.

[55] Santos, M. G., Cooray, A., Haiman, Z., Knox, L. and Ma, C.-P. (2003). Small-Scale Cosmic Microwave Background Temperature and Polarization Anisotropies Due to Patchy Reionization, *ApJ* **598**, pp. 756–766.

[56] Sasaki, S. (1994). Formation rate of bound objects in the hierarchical clustering model, *PASJ* **46**, pp. 427–430.

[57] Schaerer, D. (2002). On the properties of massive Population III stars and metal-free stellar populations, *A&A* **382**, pp. 28–42.

[58] Schaye, J., Theuns, T., Leonard, A. and Efstathiou, G. (1999). Measuring the equation of state of the intergalactic medium, *MNRAS* **310**, p. 57.

[59] Schneider, R., Salvaterra, R., Ferrara, A. and Ciardi, B. (2006). Constraints on the initial mass function of the first stars, *MNRAS* **369**, pp. 825–834.

[60] Somerville, R. S. and Primack, J. R. (1999). Semi-analytic modelling of galaxy formation: the local Universe, *MNRAS* **310**, p. 1087.

[61] Songaila, A. (2004). The Evolution of the Intergalactic Medium Transmission to Redshift 6, *AJ* **127**, pp. 2598–2603.

[62] Songaila, A. and Cowie, L. L. (2002). Approaching Reionization: The Evolution of the Ly α Forest from z=4 to z=6, *AJ* **123**, pp. 2183–2196.

[63] Spergel, D. N., Bean, R., Doré, O., Nolta, M. R., Bennett, C. L., Dunkley, J., Hinshaw, G., Jarosik, N., Komatsu, E., Page, L., Peiris, H. V., Verde, L., Halpern, M., Hill, R. S., Kogut, A., Limon, M., Meyer, S. S., Odegard, N., Tucker, G. S., Weiland, J. L., Wollack, E. and Wright, E. L. (2007). Three-Year Wilkinson Microwave Anisotropy Probe (WMAP) Observations: Implications for Cosmology, *ApJS* **170**, pp. 377–408.

[64] Stern, D., Yost, S. A., Eckart, M. E., Harrison, F. A., Helfand, D. J., Djorgovski, S. G., Malhotra, S. and Rhoads, J. E. (2005). A Galaxy at z = 6.545 and Constraints on the Epoch of Reionization, *ApJ* **619**,

pp. 12–18.

[65] Storrie-Lombardi, L. J., McMahon, R. G. and Irwin, M. J. (1996). Evolution of neutral gas at high redshift: implications for the epoch of galaxy formation, *MNRAS* **283**, p. L79.

[66] Storrie-Lombardi, L. J., McMahon, R. G., Irwin, M. J. and Hazard, C. (1994). Evolution of Lyman-limit absorption systems over the redshift range 0.40 less than Z less than 4.69, *ApJ* **427**, pp. L13–L16.

[67] Totani, T., Kawai, N., Kosugi, G., Aoki, K., Yamada, T., Iye, M., Ohta, K. and Hattori, T. (2006). Implications for Cosmic Reionization from the Optical Afterglow Spectrum of the Gamma-Ray Burst 050904 at z = 6.3, *PASJ* **58**, pp. 485–498.

[68] Willott, C. J., Delorme, P., Reylé, C., Albert, L., Bergeron, J., Crampton, D., Delfosse, X., Forveille, T., Hutchings, J. B., McLure, R. J., Omont, A. and Schade, D. (2009). Six More Quasars at Redshift 6 Discovered by the Canada-France High-z Quasar Survey, *AJ* **137**, pp. 3541–3547, doi:10.1088/0004-6256/137/3/3541.

Probing fundamental constant evolution with redshifted spectral lines

Nissim Kanekar

Ramanujan Fellow, National Centre for Radio Astrophysics,
Tata Institute of Fundamental Research, University of Pune Campus,
Ganeshkhind, Pune - 411007, India.

E-mail: *nkanekar@ncra.tifr.res.in*

Abstract: Comparisons between the redshifts of spectral lines from cosmologically-distant galaxies can be used to probe temporal changes in low-energy fundamental constants like the fine structure constant and the proton-electron mass ratio. In this article, I review the status of this field, including the best current results from techniques using this approach, the advantages and disadvantages of the different methods, and the possibilities for progress with future astronomical studies of fundamental constant evolution.

3.1 Introduction

An implicit assumption in the standard model of particle physics is that coupling constants and the ratios of particle masses do not depend on space and time. This assumption typically breaks down in higher-dimensional theories that attempt to unify the standard model with general relativity (e.g. 61, 15). A fairly generic prediction of these models is that fundamental "constants" like the fine structure constant should show space-time variation. This is one of the few low-energy predictions of theoretical extensions of the standard model and general relativity, and has hence attracted much experimental and observational attention [for recent reviews, see, e.g., Uzan [93], García-Berro *et al.* [28]].

A wide variety of methods have been used to test the possibility of changes in different fundamental constants over the last few decades. The majority of tests have focussed on the fine structure constant α (the coupling constant of the electromagnetic force) and the ratio of the proton mass to the electron mass $\mu \equiv m_p/m_e$ (which gives the relative strengths of the strong force and the electro-weak force), although constraints have also been placed on changes in the gravitational constant G and the proton gyromagnetic ratio g_p (e.g. 14, 106). Note that it is important to test the possibility of changes in all the dimensionless fundamental constants in the standard model and relativity. While most theoretical models predict simultaneous changes in different constants, the relative amplitudes of these changes are highly model-dependent and may thus serve as a discriminator between different models. For example, most unified theories expect fractional changes in μ to be far larger than those in α, by factors of $\sim 10 - 500$ (e.g. 7, 53); there are also models that yield variation only in μ, but not in α [5].

Next, the timescales of the putative changes in the constants are model-dependent, and it is also not even obvious that the changes will be monotonic in nature. This makes it difficult to directly compare results between techniques that are sensitive to changes over very different timescales (e.g. a few years versus a few Gyrs); it is hence important to search for any variation over as wide a range of timescales as possible. Further, most techniques are typically limited by (often unknown) systematics, making it critical to use entirely independent techniques in such studies, to ensure that any apparent variation is not the result of systematic effects.

In recent years, a combination of improvements in the stability of atomic and molecular clocks and other frequency standards, and the development of new types of clocks using different elemental species, have resulted in impressive progress in laboratory studies of fundamental constant evolution 24, 74, 26, 81, 83). The best current constraint on short-term variations in α is from a comparison between optical ion clocks using trapped ^{199}Hg$^+$ and ^{27}Al$^+$ ions: Rosenband *et al.* [81] obtained $\dot{\alpha}/\alpha < 4.6 \times 10^{-17}$ yr^{-1} from repeated comparisons over a one-year period [1]. Similarly, while it has been more difficult to build clocks that unambiguously probe changes in μ alone, [83] recently compared CS hyperfine and SF^6 ro-vibrational frequency standards to obtain $\dot{\mu}/\mu < 1.1 \times 10^{-13}$ per year. The primary advantage of such laboratory studies is that systematic effects are under direct control, implying that the results are likely to be more "reliable"

---

[1] Limits quoted in this review are at 2σ significance, unless explicitly stated otherwise.

than those from geological or astronomical methods. Space-based studies will yield significant improvements in the sensitivity of atomic clock probes of fundamental constant evolution in the future (e.g. 79).

The sole problem with the above atomic clock studies is that they are only sensitive to short-time variations, over timescales of a few years; geological or astronomical techniques are needed to examine the possibility of secular changes in the constants, over billions of years. The most sensitive geological results stem from measurements of relative isotopic abundances in the Oklo natural fission reactor and studies of the time variation of the β-decay rate [84, 20]. While these methods have yielded sensitive constraints on changes in α over timescales of a few Gyrs, the results tend to be model-dependent, requiring assumptions about the constancy of other parameters (e.g. 27, 29, 75). For example, Gould *et al.* [29] emphasize that their result $-0.24 \times 10^{-7} < [\Delta\alpha/\alpha] < 0.11 \times 10^{-7}$, from nuclear data from the Oklo reactor, depends critically on the assumption that α alone changes with time.

The three broad categories of astrophysical methods to probe fundamental constant evolution over cosmological timescales are, in order of increasing lookback time: (1) spectroscopic methods, using comparisons between the redshifts of different spectral lines from distant galaxies [82], (2) "CMB" methods, based on measurements of anisotropy patterns in the cosmic microwave background [33, 49], and (3) "BBN" methods, involving the measurement of the primordial abundances of light elements formed during Big Bang nucleosynthesis [51]. While the last two approaches are sensitive to changes over very large lookback times (out to a few seconds after the Big Bang, for BBN methods), the derived constraints on any putative variation depend on the values of the cosmological parameters (e.g. 80, 62). BBN studies have the additional drawback of being model-dependent, as the dependence of the nuclear binding energies on the fundamental constants is yet unknown [18]. Finally, both methods only allow a snapshot view of differences between the values of the constants today and at two epochs, $z \sim 1100$ (CMB) and $z \sim 10^{10}$ (BBN), and cannot provide information on the *evolution* of the constants at later times, which is likely to be critical in distinguishing between theoretical models. Conversely, while the first category of techniques, based on astronomical spectroscopy, probes smaller lookback times, it has the advantage of sensitivity to changes over a wide range of timescales $\sim 1 - 10$ Gyrs, via a slew of methods with different systematic effects. In this review, I will discuss the use, advantages, disadvantages, and future of such spectroscopic techniques to probe changes in three fundamental constants, the fine structure constant α, the proton-electron mass ratio $\mu \equiv m_p/m_e$, and the proton gyromagnetic ratio g_p.

3.2 Redshifted spectral lines: Background

Savedoff [82] was the first to suggest that one might test for space-time dependences in the fundamental constants by comparing the velocities of multiple spectral transitions arising in an external galaxy. This is because different spectral lines arise from different physical mechanisms (e.g. due to fine structure, hyperfine structure, molecular rotation, etc), because of which the line frequencies have different dependences on constants like α, μ and g_p. If the fundamental constants vary with space and time, the rest frequencies of different transitions too would show a space-time dependence. Specifically, a line rest frequency measured in the laboratory today would not, in general, be the same as that in an external galaxy. If such a line were used to determine the systemic velocity of the galaxy, an incorrect velocity would be obtained, due to the use of the incorrect (laboratory) line rest frequency. Further, different spectral transitions (with different dependences of the rest frequencies on the constants) would yield different estimates of the systemic velocity of the galaxy. A comparison between these velocities can then be used, in conjunction with the known dependences of the line rest frequencies on the constants, to determine the fractional change in a given constant (or combination of constants) between the space-time locations of the external galaxy and the Earth. For example, hyperfine and rotational frequencies are proportional to $g_p\alpha^2/\mu$ and $1/\mu$, respectively, implying that a comparison between hyperfine and rotational transitions is sensitive to changes in the quantity $g_p\alpha^2$.

The highest sensitivity in such studies is usually obtained with absorption lines detected in quasar spectra, arising in galaxies lying along the line of sight to the higher-redshift quasar. This is because the strength of an absorption line depends only on local physical conditions in the absorbing gas and the brightness of the background quasar, and hence does not fall off with redshift (unlike the situation for emission lines). Absorption lines are also typically much narrower than emission lines, enabling more accurate determinations of the line redshift. Most of the results discussed below are hence based on studies of lines in either absorption or stimulated emission, although it should be emphasized that studies using emission lines do provide a useful complement to absorption-based work, especially in terms of entirely different systematic effects (e.g. 4, 32)

It should also be noted that different species often arise in different regions of a gas cloud, giving rise to spectral transitions at different velocities. In fact, it is often the case that different transitions from a single species arise from different locations, due to different local excitation conditions. Such intra-cloud motions are an important source of systematic effects in studies of fundamental constant evolution. Typical intra-cloud velocities are of order ~ 10 km/s and result in systematic errors of order $(v/c) \sim 10^{-5}$ on measurements of fractional changes in different constants. Improving the sensitivity beyond this threshold requires either statistically-large samples of spectral lines, so as to average over local effects (e.g. the many-multiplet method; 21), or the use of "special" transitions, where the physics of the line mechanism causes such local effects to be negligible (e.g. the conjugate-satellites approach; 45).

3.3 Optical techniques

The profusion of strong ultraviolet (UV) atomic transitions, redshifted into the optical waveband and detectable at high signal-to-noise in absorption spectra of high-z quasars, has meant that most spectroscopic probes of fundamental constant evolution have been based on optical techniques. The three main approaches so far are (1) the alkali doublet method, (2) the many-multiplet method, and (3) the molecular hydrogen method; these are discussed in more detail below.

3.3.1 *The alkali doublet method*

A number of ions (e.g. CIV, SiIV, etc) show fine splitting between the $P_{1/2}$ and $P_{3/2}$ energy levels, and consequently have two resonance lines at nearby wavelengths, corresponding to the transitions $S_{1/2} \to P_{1/2}$ and $S_{1/2} \to P_{3/2}$. The relative separation between the line wavelengths is proportional to α^2, implying that a comparison between the line redshifts is sensitive to changes in α [82]. Until recently, this "alkali-doublet" (AD) method provided the main astronomical probe of changes in the fine structure constant (e.g. 3, 96, 37). The best current result from the AD method is $[\Delta\alpha/\alpha] < 2.6 \times 10^{-5}$, from 21 SiIV absorption doublets at $2 < z < 3$, observed with the High Resolution Echelle Spectrograph (HIRES) on the Keck telescope [73].

The AD method is nearly an order of magnitude less sensitive than the many-multiplet method (which will be discussed next), but has an important advantage in the fact that the doublet lines arise from the same gas and hence have the same shape (apart from an absolute scaling). This provides a useful test for systematic effects. Further, relative wavelength calibration across echelle orders is not usually an issue for the AD method, as the doublet lines lie at nearby wavelengths and hence, typically on the same order. However, intra-order distortions of the wavelength scale, such as those discovered by Griest *et al.* [31] for Keck-HIRES, are a source of systematic error (see below). Uncertainty in the line rest wavelengths is an important limitation for this method, and has restricted its recent application to the SiIV 1393/1402 doublet, whose wavelengths have been accurately measured in the laboratory [30]. Improved determinations of the rest wavelengths of other doublets, e.g. the CIV 1548/1550 pair, would be very useful for progress with the AD method.

3.3.2   *The many-multiplet method*

For a many-electron atom, the relativistic first-order correction to the electron energy is proportional to $(Z\alpha)^2$, where Z is the nuclear charge, but also contains a term due to many-body effects, whose sign depends on the nature of the transition [21]. The relativistic corrections are hence significantly larger for large atoms than for small ones, implying that the wavelengths of resonance transitions in large and small atoms have different dependences on α. Further, even within a single heavy atom/ion, the wavelengths of different types of transitions (e.g. $s - p$ and $d - p$ lines) have different dependences on α. The many-multiplet (MM) method is based on these effects, and involves a comparison between transitions from both different species and different multiplets in a single species to obtain a significant increase in sensitivity to fractional changes in α over the AD method. This method requires the use of large absorber samples to average out systematic effects because the different transitions are likely to have different intrinsic redshifts (e.g. different species might arise in different parts of the absorbing cloud). The MM method is also more prone to systematic effects than the AD method (e.g. 4, 72, 65). For example, the MM method uses transitions separated widely in wavelength and hence lying on different echelle orders, requiring accurate relative wavelength calibration across orders. Further, transitions from different species or different multiplets do not necessarily have the same velocity structure, again unlike the situation in the AD method.

The MM method is the only technique till date that has found statistically-significant evidence for a change in one of the fundamental constants with time: Murphy *et al.* [67] used Keck-HIRES data to obtain $[\Delta\alpha/\alpha] = [-5.4 \pm 1.2] \times 10^{-6}$ from 143 absorbers at an average redshift $\langle z \rangle = 1.75$, suggesting that α might have been smaller at earlier times [see also Webb *et al.* [98, 100] and Murphy *et al.* [68]]. This has not been confirmed by a slew of later results from the Very Large Telescope Ultraviolet Echelle Spectrograph (VLT-UVES), using the MM or similar methods [e.g. the single ion differential α measurement (SIDAM); 56], some of which are inconsistent with the Keck-HIRES result. For example, Molaro *et al.* [65] applied the SIDAM method to VLT-UVES data to obtain $[\Delta\alpha/\alpha] = (+5.7 \pm 2.7) \times 10^{-6}$ and $[\Delta\alpha/\alpha] = (-0.12 \pm 1.79) \times 10^{-6}$ from absorbers at $z \sim 1.84$ and $z \sim 1.15$, respectively [see also Levshakov *et al.* [57] and Levshakov *et al.* [58]]. Murphy *et al.* [70] argue that some of these VLT-UVES results are likely to be unreliable, notably due to the fitting of insufficient spectral components.

Note that most of the VLT-UVES results in the literature that probe changes in α are based on individual absorbers. Until very recently, the sole VLT-UVES result from a medium-sized sample (23 absorbers) is that of Srianand *et al.* [85] [based on data presented in Chand *et al.* [11]], who obtained $[\Delta\alpha/\alpha] = (-0.6 \pm 0.6) \times 10^{-6}$ at $\langle z \rangle = 1.55$, in conflict with the Keck-HIRES result of Murphy *et al.* [67]. However, Murphy *et al.* [70] showed that the χ^2 curves of Chand *et al.* [11] have large fluctuations, implying that their minimization procedure had not converged to a true minimum, and also causing them to under-estimate their errors (by a factor of $\gtrsim 3$; see also 69). While Srianand *et al.* [86] present results from a re-analysis of the VLT-UVES data of Chand *et al.* [11], and do obtain the expected larger errors, they exclude, without justification, two systems that show large values of $[\Delta\alpha/\alpha]$ from their final result. Murphy *et al.* [70] have also analysed these VLT-UVES data, but only presented results based on the original profile fits of Chand *et al.* [11]. They argued that in almost all cases, these fits are insufficient to reproduce the velocity structure in the different lines, and that a complete re-analysis is needed, using more spectral components. Given the fact that none of the present results based on the VLT-UVES sample of Chand *et al.* [11] appear reliable, I will not include any of them in the later discussion.

Very recently, Webb *et al.* [99] have applied the MM method to a VLT-UVES sample of comparable size (153 absorbers towards 60 quasars) to the Keck-HIRES sample of Murphy *et al.* [67]. Interestingly enough, the VLT-UVES sample gives $[\Delta\alpha/\alpha] = (-0.6 \pm 1.6) \times 10^{-6}$ at $\langle z \rangle < 1.8$, but $[\Delta\alpha/\alpha] = (+6.1 \pm 2.0) \times 10^{-6}$ at $\langle z \rangle > 1.8$. The high-z sample thus gives a positive value of $[\Delta\alpha/\alpha]$, in contrast to the Keck-HIRES result of Murphy *et al.* [67], albeit of a similar amplitude. Again contrary to the Keck result, the VLT-UVES sample at $z < 1.8$ is consistent with no evolution in α. Webb *et al.* [99] fit the combined VLT and Keck data with a simple spatial dipole model, and find a strong correlation, with $\sim 4.1\sigma$ significance. This result has not been examined in detail at the present time.

Excellent discussions of the systematic effects inherent in the MM (and AD) methods are given in Murphy *et al.* [72] and Murphy *et al.* [68], who find no evidence that their Keck-HIRES MM result might be affected by such systematics [but see Bahcall *et al.* [4] and Molaro *et al.* [65]]. The possibility of isotopic abundance variation with redshift is perhaps the most important amongst these systematic effects, as lines from different isotopes are blended for most species (especially Mg), and terrestrial isotopic abundances are assumed in order to determine the central line wavelength [67, 2]. In the case of Mg, higher ^{25,26}Mg fractional abundances in the $z < 1.8$ sample of Murphy *et al.* [67] could yield the observed negative $[\Delta\alpha/\alpha]$ obtained here. Interestingly, Molaro *et al.* [65] find that the absorber sample based on MgII lines dominates the results of Murphy *et al.* [67]: the sub-sample using MgII lines gives a β-trimmed mean of $[\Delta\alpha/\alpha] = (-0.48 \pm 0.12) \times 10^{-5}$, while the sub-sample without these lines gives $[\Delta\alpha/\alpha] = (-0.11 \pm 0.17) \times 10^{-5}$ [65]. Further, the isotopic structure is only known for a few of the transitions used in the MM analysis [72]. Unfortunately, it is difficult to resolve this issue by direct observations, given the low spectral resolution of optical spectra, while indirect arguments depend on the details of models of galactic chemical evolution [2, 23].

An important new source of systematic error in Keck-HIRES data was recently discovered by Griest *et al.* [31], by comparing wavelength solutions from the standard ThAr wavelength calibration (where the quasar and ThAr lamp have different light paths and illuminate the slit differently) with an absolute wavelength calibration obtained by placing an iodine cell in the quasar light path. Griest *et al.* [31] found evidence for intra-order distortions in the ThAr wavelength calibration solutions between ~ 5000Å and ~ 6200Å, with the errors lowest at the edges of each order, and increasing to a maximum of ~ 1000 m/s near the order centres. Griest *et al.* [31] also found the ThAr wavelength solution to drift with time, with offsets of ~ 500 m/s within a night, and of ~ 2 km/s over different nights.

Wavelength errors of similar amplitude, as well as drifts of the wavelength scale with time, have been seen earlier in Keck-HIRES spectra (e.g. 89). The intra-order distortions are especially worrisome for the MM and AD methods, as they would give rise to velocity offsets between different transitions. Murphy *et al.* [71] argue that the distortions should average out over large absorber and line samples, as the different absorber redshifts would cause various lines to lie at different locations in each echelle order, or indeed, in different echelle orders. They also used a simple saw-tooth model for the intra-order distortions seen by Griest *et al.* [31], extrapolating this to wavelengths not covered by the iodine cell calibration, and found very little effect of the distortions on the results for $[\Delta\alpha/\alpha]$ in individual absorbers. However, it should be emphasized that the origin of the distortions and drifts, and their dependence on physical conditions during the observations, are entirely unknown. This is clearly an important possible source of systematic errors for the result of Murphy *et al.* [67]. Similar distortions (with amplitudes of a few hundred m/s) have also been later seen in VLT-UVES data, by comparing multiple FeII lines from a single absorber [10] and using an iodine cell [102].

With regard to the results of Webb *et al.* [99], it should perhaps be emphasized that the wavelength calibration issues found by Griest *et al.* [31] and Whitmore *et al.* [102] certainly affect their data, providing an added source of systematic error. Seven systems of their sample were observed with both the Keck-HIRES and VLT-UVES spectrographs, allowing a comparison between the two wavelength scales. For six of the systems, Webb *et al.* [99] fit the observed velocity differences between the HIRES and UVES wavelength scales with a linear function of wavelength; applying this function to either dataset reduced the statistical significance of the dipole to 3.1σ. Even more worrying, the seventh system was found to show significantly more complex calibration problems. Webb *et al.* [99] attempted to model these with a non-linear transformation, applying the model to their data in a Monte-Carlo manner. This caused the statistical significance of their result to reduce to 2.2σ. While Webb et al. argue that this is an over-estimate of a systematic effect of this type, it appears clear that the data contain wavelength calibration uncertainties that are not well understood. It is thus probably reasonable at present to be sceptical of the claimed detection of a spatial dipole.

3.3.3   *Molecular hydrogen lines*

Molecular hydrogen (H_2), the most abundant molecule in the Universe, has a number of UV ro-vibrational transitions whose wavelengths have different dependences on the reduced molecular mass. Comparisons between the H_2 line redshifts can hence be used to probe changes in the proton-electron mass ratio μ [90, 95]. Unfortunately, it is difficult to detect these lines as they are weak and, for redshifted systems, lie in the Lyman-α forest. Only about a dozen redshifted H_2 absorbers are currently known, most as a result of deep VLT-UVES searches in high-z damped Lyman-α systems (e.g. 54). So far, only four systems have yielded useful constraints on changes in μ (e.g. 95, 78, 50, 91, 60, 101). The best present result is that of King *et al.* [50] who used VLT-UVES data on three absorbers at $z \sim 2.6 - 3$ to obtain $[\Delta\mu/\mu] < 6.0 \times 10^{-6}$. However, the highest-sensitivity spectrum of King *et al.* [50], from the $z \sim 2.811$ absorber towards PKS 0528−25, shows a complicated velocity structure in the H_2 lines, implying that the result might be affected by the presence of unresolved spectral components. King *et al.* [50] also carry out a simultaneous fit to both the H_2 lines and all the Lyman-α forest transitions in their vicinity, rather than excluding possible blends with Lyman-α forest interlopers (e.g. 78, 91, 101). While this allowed King et al. to retain a far larger number of H_2 lines for their analysis, it is unclear what effect it might have on their results. The quoted errors in the analysis (and in some of the other analyses in the literature) also do not include the intra-order distortions affecting VLT-UVES data that use the standard ThAr wavelength calibration [10, 102]; this is discussed in detail by Wendt and Molaro [101].

3.4   "Radio" techniques

The methods discussed in the preceding section are all based exclusively on optical spectroscopy, and are all affected by the intra-order distortions of the wavelength scale on using the ThAr wavelength calibration scheme [31]. Other sources of systematic error at optical wavebands include line blending at the low velocity resolution ($\sim 6 - 8$ km/s) of current quasar spectra, line interlopers, saturation effects, etc (e.g. 68). The fact that redshifted H_2 lines fall in the Lyman-α forest complicates their use in measuring $[\Delta\mu/\mu]$, while the MM method is affected by the issue of unknown isotopic abundances in high-z galaxies.

Techniques based on radio spectroscopy provide independent approaches to probe changes in α, μ and g_p, with the immediate benefit of very different systematic effects from those in the optical regime. A number of different physical mechanisms (e.g. hyperfine splitting, molecular rotation, Lambda-doubling, etc), or combinations thereof, give rise to radio spectral lines, and the line frequencies hence can have very different dependences on the fundamental constants. Comparisons between the redshifts of radio lines (or between radio and optical lines; 107) can thus be used to probe changes in the fundamental constants. Note that frequency calibration is not an issue at radio wavelengths as the frequency scale is set by masers and local oscillators (allowing wavelength calibration to accuracies of $\lesssim 10$ m/s), while the excellent velocity resolution ($\lesssim 1$ km/s) available at radio wavebands alleviates problems with line blending. In this section, I will first discuss a "hybrid" method, based on comparing radio and optical lines, before considering techniques based entirely on radio spectroscopy.

3.4.1 *Radio-optical comparisons*

The frequencies of UV resonance dipole transitions are independent of α, μ and g_p to zeroth order, except for the dependence on α through the Rydberg constant R. On the other hand, radio line frequencies are, in general, proportional to $F(\alpha, \mu, g_p) \times R$, where F is some function of α, μ and g_p. This means that comparisons between the redshifts of UV resonance lines and radio lines from a single system are sensitive to changes in $F(\alpha, \mu, g_p)$. Wolfe *et al.* [107] were the first to suggest the use of the HI 21cm line for this purpose, comparing HI 21cm and metal-line (MgII) absorption redshifts in a $z \sim 0.524$ absorber to constrain changes in the quantity $X \equiv g_p \alpha^2 / \mu$. A similar comparison between the redshifts of the HI 21cm line and low-ionization metal lines was used by Tzanavaris *et al.* [92] to obtain $[\Delta X/X] < 2 \times 10^{-5}$ from nine redshifted HI 21cm absorbers at $0.23 < z < 2.35$, although this involved the assumption that the deepest HI 21cm and metal-line absorption components arise in the same gas (which is by no means necessary for complex profiles; e.g. 48).

Absorbers with a single (or dominant) spectral component in both HI 21cm and resonance lines are likely to yield the most reliable results from the resonance-hyperfine line comparison. The best resonance lines for this purpose are those arising from *neutral* atomic species (e.g., CI, MgI, etc), as these are most likely to be physically associated with the neutral hydrogen. Of these, the CI multiplets are likely to be best-suited for this method, as they typically arise in the cold gas that also gives rise to HI 21cm absorp-

tion (e.g. Jenkins and Tripp 38, Srianand *et al.* 87). CI and HI also have
similar ionization potentials, 11.3 eV and 13.6 eV, respectively; cf. MgI,
which is more easily detectable than CI in high-z absorbers, has an ionization potential of 7.6 eV, as well as a high dielectronic recombination rate
that can yield significant MgI absorption in warm, ionized gas [76]. Unfortunately, CI and HI 21cm lines have both so far been detected in only three
redshifted absorbers: the best result from this method is that of Kanekar
et al. [47] who obtained $[\Delta X/X] = [+6.8 \pm 1.0(stat.) \pm 6.7(syst.)] \times 10^{-6}$,
where $X \equiv g_p \alpha^2/\mu$, based on two absorbers at $\langle z \rangle = 1.46$, with narrow,
single-component profiles in both CI and HI 21cm lines. This result is inconsistent with the lower value of α found by Murphy *et al.* [67] in their
Keck-HIRES sample, unless changes in g_p are larger than those in α and μ
[47].

Note that the above CI-HI comparison suffers from all the wavelength
calibration issues that affect optical spectroscopy. In fact, it is worse than
the MM methods in this regard, because it requires accurate *absolute* wavelength calibration of the optical spectra (unlike the MM method, which only
needs accurate *relative* wavelength calibration). Conversely, the CI-HI comparison uses the zeroth-order dependences of the resonance and hyperfine
frequencies on the constants, while the MM method relies on a first-order
effect (the line wavelengths in the MM method have the same zeroth-order
dependence on α). As a result, systematic effects (e.g. due to wavelength
mis-calibration) are less important by an order of magnitude in the hyper-
fine/resonance comparison than in the MM analysis [47].

3.4.2 *Radio comparisons*

As noted above, frequency calibration is not a serious problem in techniques probing fundamental constant evolution that rely entirely on radio transitions. The HI 21cm hyperfine transition has been used in most
such comparisons so far, in conjunction with molecular lines from different
species (e.g. CO, OH, etc). For example, comparisons between the redshifts of HI 21cm and molecular rotational transitions (e.g. CO, HCO$^+$,
etc) are sensitive to changes in $Y \equiv g_p \alpha^2$. Unfortunately, there are only
four redshifted systems with detections of both HI 21cm and rotational
molecular absorption (e.g. 103, 105, 8, 12), of which only two are suitable to probe changes in the constants. Carilli *et al.* [9] used these to
obtain $[\Delta Y/Y] < 3.4 \times 10^{-5}$ from $z \sim 0.25$ and $z \sim 0.685$, conservatively
assuming that differences between the HI 21cm and millimetre sightlines

could yield local velocity offsets of ~ 10 km/s. Similarly, comparisons between the HI 21cm and "main" OH 18cm lines are sensitive to changes in $Z \equiv g_p[\mu\alpha^2]^{1.57}$ (13; see also 16). This method was applied by Kanekar *et al.* [42] to HI 21cm and OH 18cm lines from two gravitational lenses to obtain $[\Delta Z/Z] < 2.1 \times 10^{-5}$ from $\langle z \rangle \sim 0.7$, where the errors are dominated by the assumed local velocity offsets of ~ 3 km/s between the HI and OH components. Other "radio" methods of this kind include comparisons between CII-158μm and CO lines [59], between, and within, CH and OH lines [13, 16, 43, 52], between inversion and rotational lines (discussed in detail below; 25), etc. An important source of systematic effects in all these techniques is that sightlines in the different transitions may probe different velocity structures in the absorbing gas. For example, large velocity offsets (~ 15 km/s) have been observed between the HI 21cm and HCO$^+$ redshifts at $z \sim 0.674$ towards B1504+377, probably due to small-scale structure in the absorbing gas [104, 8, 44]. This is especially important for comparisons based on small samples, where such local velocity offsets do not average out.

3.4.3 *Ammonia inversion transitions*

The inversion lines in the ammonia (NH$_3$) molecule arise due to the tunneling of the three hydrogen atoms through a potential barrier, with the line frequency having a strong dependence on the tunneling probability, and thence, on the reduced mass of the system. This fact was used by Flambaum and Kozlov [25] to demonstrate that a comparison between inversion and rotational transitions is highly sensitive to changes in the proton-electron mass ratio μ. The NH$_3$ lines have so far been detected in only two cosmologically-distant absorbers, the $z \sim 0.685$ and $z \sim 0.886$ gravitational lenses towards B0218+357 and B1830$-$210 [35, 34]. While the system towards B1830$-$210 has been used for such studies (e.g. 63, 36), some of its NH$_3$ lines (especially the stronger 1-1, 2-2, and 3-3 transitions) are affected by satellite radio frequency interference, rendering their redshift estimates unreliable. The absorber towards B0218+357 also has narrow line profiles, and has hence yielded the most stringent constraints on changes in μ [25, 66, 41]. The tightest limit from this absorber is that of Kanekar [41] who compared the redshift of the NH$_3$ (1,1) inversion lines with those of the CS 1-0 and H$_2$CO 0_{00}-1_{01} rotational transitions to obtain $[\Delta\mu/\mu] < 3.6 \times 10^{-7}$ from $z \sim 0.685$, at 3σ significance. A comparison between the CS and H$_2$CO rotational lines was used to obtain an estimate of the velocity dispersion in the absorbing clouds, with the redshifts of the CS and H_2CO lines found to agree within ~ 68 m/s, significantly lower than the statistical errors in the comparison between NH_3 and CS lines. Unknown systematic effects include (1) time variability in the morphology of the background source, due to which the sightlines through the molecular clouds might be different at different epochs, and (2) velocity offsets between nitrogen-bearing and carbon-bearing molecular species. At present, this is the strongest constraint on fractional changes in any constant from any astronomical technique.

3.5 "Conjugate" Satellite OH lines

"Conjugate" satellite OH lines provide perhaps the best astronomical technique to probe the possibility of fundamental constant evolution. The ground-state satellite OH 18cm lines (i.e. lines with unit change in the total angular momentum, $\Delta F = 1$) are said to be conjugate when the two lines have the same shape, but with one line in emission and the other in absorption. This arises due to an inversion of the level populations within the OH ground state [22, 94], and is discussed in detail in Kanekar [40]. Only two redshifted conjugate satellite OH systems are currently known, at $z \sim 0.247$ towards PKS 1413+135 [45, 17] and $z \sim 0.765$ towards PMN J0134−0931 [42]. Deep recent observations of the satellite OH lines of the former system with the Westerbork Synthesis Radio Telescope (WSRT) and the Arecibo telescope have yielded tentative evidence for changes in one or more of α, μ and g_p [46]. The combined result from the two telescopes is $[\Delta G/G] = (-1.18 \pm 0.46) \times 10^{-5}$ (weighted mean), where $G \equiv g_p \left[\mu \alpha^2 \right]^{1.85}$, suggesting (with 2.6σ significance, or at 99.1% confidence) smaller values of α, μ, and/or g_p at $z \sim 0.247$, i.e. at a lookback time of ~ 2.9 Gyrs [46]. Applying the same technique to a nearby conjugate satellite system, Cen.A [94] yielded the expected null result, with $[\Delta G/G] < 1.16 \times 10^{-5}$, from $z \sim 0.0018$.

The strength of the conjugate-satellites technique stems from the fact that the conjugate behaviour *guarantees that the satellite OH lines arise from the same gas.* Such systems are ideal to probe changes in α, μ and g_p from the source redshift to the present epoch, as local velocity offsets between the lines are ruled out by the inversion mechanism. Any measured difference between the line redshifts must arise due to changes in one (or more) of α, μ and g_p [45]. The technique also contains a stringent test of its own applicability, as the shapes of the two lines must agree if they arise in the same gas. Further, the velocity offset between the lines can be determined from a cross-correlation analysis, as the shapes of the lines are the same; it is not necessary to model the line profiles with multiple spectral components of an assumed shape, which can, especially for complex profiles, itself affect the result. The use of accurate masers and local oscillators to set the radio frequency scale means that frequency calibration is also not an issue here, while the satellite line frequencies have been measured to high accuracy in the laboratory (fractional accuracy better than 10^{-8}; 55). Finally, the nearest isotopic OH transitions are more than 50 MHz away and this part of the radio spectrum has very few other known astronomical transitions; line interlopers are thus unlikely to be a contaminant for this method, unlike the situation in the optical regime. Overall, systematic effects appear to be far less important for the conjugate-satellites method than for the other techniques. Conversely, the main drawback of the conjugate-satellites technique is the fact that, unlike the MM, AD, and H_2 methods, it cannot be used to directly measure changes in an individual constant, but is sensitive to changes in a combination of α, μ and g_p.

3.6   Results from the different techniques

Figure 3.1 shows a comparison between results from the best radio and optical techniques based on astronomical spectroscopy. Of course, the radio methods typically probe combinations of α, μ and g_p, making it difficult to directly compare results from the different techniques without additional assumptions. Following Kanekar [40], the figure summarizes the best current results for two limiting cases, $[\Delta\alpha/\alpha] >> [\Delta\mu/\mu]$ and $[\Delta\alpha/\alpha] << [\Delta\mu/\mu]$, assuming that fractional changes in g_p are much smaller than those in α and μ (e.g. 53). Results from the alkali doublet, many-multiplet, SIDAM, HI 21cm vs. CI, HI 21cm vs. OH, NH₃ vs. rotational, conjugate-satellites, and H_2 methods are plotted in the figure [73, 67, 50, 42, 65, 47, 91, 101, 60, 41]. It is apparent that the best results from the conjugate-satellites, MM and SIDAM methods have similar sensitivities to changes in α, although at very different lookback times, and with different systematic effects, while the NH₃ technique is the most sensitive to changes in μ. The Keck-HIRES MM result of Murphy *et al.* [67] remains the only one with statistically-significant evidence for changes in one of the constants, although tentative evidence for changes in one or more of α, μ,

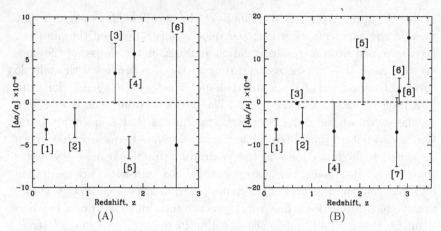

Fig. 3.1 The best current results from techniques probing fundamental constant evolution with redshifted spectral lines. The left panel [A] shows estimates for $[\Delta\alpha/\alpha]$ as a function of redshift, from comparisons between [1] conjugate satellite OH lines [46], [2] HI 21cm and OH lines (Kanekar et al, in prep.), [3] HI 21cm and CI lines [47], [4] optical lines, with the MM method [67], [5] FeII lines using the SIDAM method [65], and [6] SiIV alkali-doublet lines [73]. Panel [B] shows similar results for $[\Delta\mu/\mu]$, using comparisons between [1] conjugate satellite OH lines [46], [2] HI 21cm and OH lines (Kanekar et al, in prep.), [3] NH$_3$ and rotational lines [41], [4] HI 21cm and CI lines [47], and H$_2$ lines ([5] 60; [6] 91; [7] 50, and [8] 101. The assumptions $[\Delta\alpha/\alpha] \gg [\Delta\mu/\mu], [\Delta g_p/g_p]$ (in [A]) and $[\Delta g_p/g_p], [\Delta\alpha/\alpha] \ll [\Delta\mu/\mu]$ (in [B]) apply to the first three for $[\Delta\alpha/\alpha]$ and [1], [2], and [4] for $[\Delta\mu/\mu]$.

and g_p has been found by Kanekar *et al.* [46], using the conjugate-satellites method.

Figure 3.1 also emphasizes the separation between the redshift ranges at which the radio and optical techniques have been used. The optical techniques are all based on UV transitions that only fall into optical wavebands (say, $\gtrsim 3200$Å) for absorbers beyond a certain redshift. For example, the H$_2$ technique can only be used at $z \gtrsim 2$, the SiIV AD technique at $z \gtrsim 1.3$ and the SIDAM method (using the FeIIλ1608 line; 77) at $z \gtrsim 1$. Similarly, only a handful of lines from singly-ionized species are redshifted above 3200Å from $z \lesssim 0.6$, implying that the MM method also works best at higher redshifts ($z \gtrsim 0.8$). On the other hand, the lack of known radio molecular absorbers at $z > 0.9$ limits the current application of the radio methods to relatively low redshifts. Present radio and optical techniques thus play complementary roles in studies of fundamental constant evolution, with the best low-z measurements coming from radio wavebands and the best high-z

estimates from the optical regime.

3.7 Future studies

The present limitations of the optical and radio techniques are very different in nature. The optical methods are affected by issues like line blending, relative calibration of different echelle orders, unknown relative isotopic abundances at high redshifts, etc. Further, all the "optical" methods are based on UV transitions, making it difficult to obtain an estimate of local systematic effects by applying the techniques to Galactic lines of sight. Conversely, the shortage of HI 21cm and radio molecular absorbers at high redshifts is the biggest drawback of the radio-based techniques; this means that local velocity offsets may not be averaged out and are an important source of systematics (this does not affect the conjugate-satellites method).

Over the next decade, many of the above issues will be addressed by new telescopes and associated instrumentation at both radio and optical wavebands. The next generation of 30m-class optical telescopes and new spectrographs will yield higher sensitivity, improved spectral resolution (alleviating problems with line blends, especially useful for the H_2 method), and far better wavelength calibration. For example, laser frequency combs offer the possibility of optical wavelength calibration to accuracies of ~ 1 cm/s [88], although much work remains to be done on systematic effects associated with the spectrograph and detector systems. The improved sensitivity and resolution should result in a significant increase in the number of both redshifted H_2 absorbers and alkali-doublet pairs suitable for studies of changes in μ and α. Note, however, that even the planned spectrograph CODEX, with a resolving power of $R \sim 150000$ (e.g. 64), will be unable to resolve out the isotopic structure in various species (e.g. Mg, where the different isotopic transitions are separated by only ~ 0.85 km/s and ~ 0.4 km/s for MgII$\lambda 2803$ and MgI$\lambda 2853$, respectively; 73). Laboratory or theoretical determinations of the isotopic structure of different species are hence important, to allow independent applications of the MM or SIDAM methods to different species with different isotopic structure (e.g. 19, 6). Note that, although the improved sensitivity should allow statistical errors of $[\Delta \alpha / \alpha] \sim 10^{-7}$ with the MM method, unknown relative isotopic abundances in the high-z absorbers are likely to remain a source of systematic effects, unless one can use multiple species to circumvent the problem. Extensions of the MM method (e.g. the SIDAM method; 56) are also likely to be of much interest, especially if such extensions can be applied to species with widely separated isotopic transitions. In the case of the H_2 method, higher-order corrections to the Born-Oppenheimer approximation are likely to be necessary to attain sensitivities of $[\Delta\mu/\mu] \sim 10^{-7}$ (e.g. 78). It will also be important to more accurately determine the laboratory wavelengths of both the H_2 lines and the transitions used in the MM and AD methods, some of which are only known to a fractional accuracy of $\sim 10^{-6}$ (e.g. 1). Unfortunately, it is unlikely that it will be possible to examine systematic effects affecting these methods through Galactic studies, as this would require space-based ultraviolet spectroscopy.

On the radio front, increasing the number of redshifted systems detected in atomic and molecular radio lines is critically needed for improvements in radio studies of fundamental constant evolution. This too should be feasible with upcoming telescopes and instrumentation. The wide-band receivers and correlators of the Atacama Large Millimeter Array and the Expanded Very Large Array (EVLA) will, for the first time, allow "blind" surveys for redshifted absorption in the strong mm-wave rotational transitions towards a statistically-large number of background sources. This should yield sizeable samples of high-z absorbers in these transitions, which can then be followed up in the OH, NH_3, and HI 21cm lines (note that two of the five known radio molecular absorbers show conjugate-satellite behaviour). The new EVLA 1.4 GHz receivers will allow "blind" surveys for redshifted HI 21cm and OH 18cm absorption out to $z_{HI} \sim 0.5$ and $z_{OH} \sim 0.7$, with every L-band continuum observation automatically yielding such an absorption survey towards all background sources in the field of view. Similarly, wide field-of-view surveys with the Australian SKA Pathfinder array [39] or APERTIF on the WSRT [97] should yield hundreds of new HI 21cm absorbers at $0 < z < 1$ over the next five years (besides new OH 18cm absorbers and conjugate-satellite OH systems). It should then be possible to average out local velocity offsets and different kinematic structures to obtain reliable results when comparing redshifts between different radio lines, or between radio and optical lines. It should also be possible to achieve 1σ sensitivities of $[\Delta\alpha/\alpha]$, $[\Delta\mu/\mu] \sim few \times 10^{-7}$ with deep ($>> 100$ hour) integrations on the $z \sim 0.247$ conjugate system in PKS 1413+135 with existing telescopes over the next few years. In the long-term future, the Square Kilometer Array will be able to detect fractional changes of $[\Delta\alpha/\alpha] \sim 10^{-7}$ in both known, and any newly-detected, conjugate systems. Finally, it is important to recognize that very few molecules have been examined for their use as probes of fundamental constant evolution; theoretical studies in this area are hence of much importance, both to obtain new techniques with independent systematic effects, and to find transitions with strong dependences on different constants (e.g. 52).

3.8   Summary

Optical and radio techniques have played complementary roles in extending studies of fundamental constant evolution to large lookback times. Optical approaches have provided high sensitivity at high redshifts, $z \sim 0.8 - 3.0$, while radio approaches have been critical to study the late-time ($z \lesssim 0.8$) behaviour. The best present results, from a variety of methods, have 2σ sensitivities of $[\Delta\alpha/\alpha]$, $[\Delta\mu/\mu] \sim few \times 10^{-6}$ at $z \sim 0.25 - 2.8$, an improvement of nearly two orders of magnitude over the last decade. At present, it appears fair to say that the Keck-HIRES MM result, $[\Delta\alpha/\alpha] = [-0.57 \pm 0.11] \times 10^{-5}$ at $\langle z \rangle = 1.75$, which suggests a smaller value of α at high redshifts [67] has been neither contradicted nor confirmed by any other study. Recently, the conjugate-satellites method also found tentative evidence (with $\sim 2.6\sigma$ significance) for smaller values of α, μ and/or g_p at $z \sim 0.247$ [46]. It is of critical importance to confirm or deny these results, with either observations at independent telescopes, or independent techniques.

It also appears that significant progress is likely to be possible over the next decade in astronomical studies of fundamental constant evolution, reaching a sensitivity to fractional changes in α and μ comparable to that (in $[\Delta\alpha/\alpha]$) from the Oklo nuclear reactor, with far fewer assumptions, over a larger lookback time, and via multiple techniques. The lack of apparent systematic effects in the conjugate-satellites method suggests that it is likely to be the most reliable probe of fundamental constant evolution, unless the issue of unknown isotopic abundances in the MM method can be resolved. However, the conjugate-satellites method suffers from the drawback of being sensitive to changes in a combination of α, μ and g_p, implying that other approaches will be necessary to measure changes in the individual constants. In any event, it remains important that multiple independent techniques be used, both to ensure that the results are not dominated by systematic effects in any given approach and to probe changes in multiple constants over a wide range of redshifts.

3.9   Acknowledgments

It is a pleasure to thank Jayaram Chengalur, Carlos Martins, Michael Murphy, and Xavier Prochaska for stimulating discussions on fundamental constant evolution, and T. Padmanabhan for many illuminating conversations on science over the last fifteen years. I acknowledge support from a Ramanujan Fellowship from the Department of Science and Technology.

References

[1] Aldenius, M. (2009). *Physica Scripta* **134**, p. 014008.

[2] Ashenfelter, T. P., Mathews, G. J. and Olive, K. A. (2004). *ApJ* **615**, p. 82.

[3] Bahcall, J. N., Sargent, W. L. W. and Schmidt, M. (1967). *ApJ* **149**, p. L11.

[4] Bahcall, J. N., Steinhardt, C. L. and Schlegel, D. (2004). *ApJ* **600**, p. 520.

[5] Barrow, J. D. and Magueijo, J. (2005). *Phys. Rev. D* **72**, p. 043521.

[6] Berengut, J. C., Flambaum, V. V. and Kozlov, M. G. (2006). *Phys. Rev. A* **73**, p. 012504.

[7] Calmet, X. and Fritzsch, H. (2002). *Eur. Phys. Jour. C* **24**, p. 639.

[8] Carilli, C. L., Menten, K. M., Reid, M. J. and Rupen, M. P. (1997). *ApJ* **474**, p. L89.

[9] Carilli, C. L., Menten, K. M., Stocke, J. T., Perlman, E., Vermeulen, R., Briggs, F., de Bruyn, A. G., Conway, J. and Moore, C. P. (2000). *Phys. Rev. Lett.* **85**, p. 5511.

[10] Centurion, M., Molaro, P. and Levshakov, S. (2009). *MmSAI, proc. of IAU JD 9, P. Molaro & E. Vangoni eds.* **80**.

[11] Chand, H., Srianand, R., Petitjean, P. and Aracil, B. (2004). *A&A* **417**, p. 853.

[12] Chengalur, J. N., de Bruyn, A. G. and Narasimha, D. (1999). *A&A* **343**, p. L79.

[13] Chengalur, J. N. and Kanekar, N. (2003). *Phys. Rev. Lett.* **91**, p. 241302.

[14] Copi, C. J., Davis, A. N. and Krauss, L. M. (2004). *Phys. Rev. Lett.* **92**, p. 171301.

[15] Damour, T. and Polyakov, A. M. (1994). *Nucl. Phys. B* **423**, p. 532.

[16] Darling, J. (2003). *Phys. Rev. Lett.* **91**, p. 011301.

[17] Darling, J. (2004). *ApJ* **612**, p. 58.

[18] Dent, T., Stern, S. and Wetterich, C. (2007). *Phys. Rev. D* **76**, p. 063513.

[19] Drullinger, R. E., Wineland, D. J. and Bergquist, J. C. (1980). *Appl. Phys.* **22**, p. 365.

[20] Dyson, F. J. (1967). *Phys. Rev. Lett.* **19**, p. 1291.

[21] Dzuba, V. A., Flambaum, V. V. and Webb, J. K. (1999). *Phys. Rev. Lett.* **82**, p. 888.

[22] Elitzur, M. (1992). *Astronomical masers* (Kluwer Academic, Dordrect, NL).

[23] Fenner, Y., Murphy, M. T. and Gibson, B. K. (2005). *MNRAS* **358**, p. 468.

[24] Fischer, M., Kolachevsky, N., Zimmermann, M., Holzwarth, R., Udem, T., Hänsch, T. W., Abgrall, M., Grünert, J., Maksimovic, I., Bize, S., Marion, H., Santos, F. P., Lemonde, P., Santarelli, G., Laurent, P., Clairon, A., Salomon, C., Haas, M., Jentschura, U. D. and Keitel, C. H. (2004). *Phys. Rev. Lett.* **92**, p. 230802.

[25] Flambaum, V. V. and Kozlov, M. G. (2007). *Phys. Rev. Lett.* **98**, p. 240801.

[26] Fortier, T. M., Ashby, N., Bergquist, J. C., Delaney, M. J., Diddams, S. A., Heavner, T. P., Hollberg, L., Itano, W. M., Jefferts, S. R., Kim, K., Levi, F., Lorini, L., Oskay, W. H., Parker, T. E., Shirley, J. and Stalnaker, J. E. (2007). *Phys. Rev. Lett* **98**, p. 070801.

[27] Fujii, Y. and Iwamoto, A. (2003). *Phys. Rev. Lett.* **91**, 26, p. 261101.

[28] García-Berro, E., Isern, J. and Kubyshin, Y. A. (2007). *A&AR* **14**, p. 113.

[29] Gould, C. R., Sharapov, E. I. and Lamoreaux, S. K. (2006). *Phys. Rev. C* **74**, p. 024607.

[30] Griesmann, U. and Kling, R. (2000). *ApJ* **536**, p. L113.

[31] Griest, K., Whitmore, J. B., Wolfe, A. M., Prochaska, J. X., Howk, J. C. and Marcy, G. W. (2010). *ApJ* **708**, p. 158.

[32] Grupe, D., Pradhan, A. K. and Frank, S. (2005). *AJ* **130**, p. 355.

[33] Hannestad, S. (1999). *Phys. Rev. D* **60**, p. 023515.

[34] Henkel, C., Braatz, J. A., Menten, K. M. and Ott, J. (2008). *A&A* **485**, p. 451.

[35] Henkel, C., Jethava, N., Kraus, A., Menten, K. M., Carilli, C. L., Grasshoff, M., Lubowich, D. and Reid, M. J. (2005). *A&A* **440**, p. 893.

[36] Henkel, C., Menten, K. M., Murphy, M. T., Jethava, N., Flambaum,

V. V., Braatz, J. A., Muller, S., Ott, J. and Mao, R. Q. (2009). *A&A* **500**, p. 725.

[37] Ivanchik, A. V., Potekhin, A. Y. and Varshalovich, D. A. (1999). *A&A* **343**, p. 439.

[38] Jenkins, E. B. and Tripp, T. M. (2001). *ApJS* **137**, p. 297.

[39] Johnston, S. and et al. (2007). *PASA* **24**, p. 174.

[40] Kanekar, N. (2008). *Mod. Phys. Lett. A* **23**, p. 2711.

[41] Kanekar, N. (2010). *Phys. Rev. Lett. (submitted)* .

[42] Kanekar, N., Carilli, C. L., Langston, G. I., Rocha, G., Combes, F., Subrahmanyan, R., Stocke, J. T., Menten, K. M., Briggs, F. H. and Wiklind, T. (2005). *Phys. Rev. Lett.* **95**, p. 261301.

[43] Kanekar, N. and Chengalur, J. N. (2004). *MNRAS* **350**, p. L17.

[44] Kanekar, N. and Chengalur, J. N. (2008). *MNRAS* **384**, p. L6.

[45] Kanekar, N., Chengalur, J. N. and Ghosh, T. (2004). *Phys. Rev. Lett.* **93**, p. 051302.

[46] Kanekar, N., Chengalur, J. N. and Ghosh, T. (2010). *ApJ* **716**, p. L23.

[47] Kanekar, N., Prochaska, J. X., Ellison, S. L. and Chengalur, J. N. (2010). *ApJ* **712**, p. L148.

[48] Kanekar, N., Subrahmanyan, R., Ellison, S. L., Lane, W. M. and Chengalur, J. N. (2006). *MNRAS* **370**, p. L46.

[49] Kaplinghat, M., Scherrer, R. J. and Turner, M. S. (1999). *Phys. Rev. D* **60**, p. 023516.

[50] King, J. A., Webb, J. K., Murphy, M. T. and Carswell, R. F. (2008). *Phys. Rev. Lett.* **101**, p. 251304.

[51] Kolb, E. W., Perry, M. J. and Walker, T. P. (1986). *Phys. Rev. D* **33**, p. 869.

[52] Kozlov, M. G. (2009). *Phys. Rev. A* **80**, p. 022118.

[53] Langacker, P. G., Segré, G. and Strassler, M. J. (2002). *Phys. Lett. B* **528**, p. 121.

[54] Ledoux, C., Petitjean, P. and Srianand, R. (2003). *MNRAS* **346**, p. 209.

[55] Lev, B. L., Meyer, E. R., Hudson, E. R., Sawyer, B. C., Bohn, J. L. and Ye, J. (2006). *Phys. Rev. A* **74**, p. 061402.

[56] Levshakov, S. A., Centurión, M., Molaro, P. and D'Odorico, S. (2005). *A&A* **434**, p. 827.

[57] Levshakov, S. A., Centurión, M., Molaro, P., D'Odorico, S., Reimers, D., Quast, R. and Pollmann, M. (2006). *A&A* **449**, p. 879.

[58] Levshakov, S. A., Molaro, P., Lopez, S., D'Odorico, S., Centurión,

M., Bonifacio, P., Agafonova, I. I. and Reimers, D. (2007). *A&A* **466**, p. 1077.

[59] Levshakov, S. A., Reimers, D., Kozlov, M. G., Porsev, S. G. and Molaro, P. (2008). *A&A* **479**, p. 719.

[60] Malec, A. L., Buning, R., Murphy, M. T., Milutinovic, N., Ellison, S. L., Prochaska, J. X., Kaper, L., Tumlinson, J., Carswell, R. F. and Ubachs, W. (2010). *MNRAS* **403**, p. 1541.

[61] Marciano, W. J. (1984). *Phys. Rev. Lett.* **52**, p. 489.

[62] Menegoni, E., Galli, S., Bartlett, J. G., Martins, C. J. A. P. and Melchiorri, A. (2009). *Phys. Rev. D* **80**, p. 087302.

[63] Menten, K. M., Güsten, R., Leurini, S., Thorwirth, S., Henkel, C., Klein, B., Carilli, C. L. and Reid, M. J. (2008). *A&A* **492**, p. 725.

[64] Molaro, P., Murphy, M. T. and Levshakov, S. A. (2006). in P. Whitelock et al. (ed.), *The Scientific Requirements for Extremely Large Telescopes* (Cambridge University Press, Cambridge), p. 198.

[65] Molaro, P., Reimers, D., Agafonova, I. I. and Levshakov, S. A. (2008). *European Physical Journal Special Topics* **163**, p. 173.

[66] Murphy, M. T., Flambaum, V. V., Muller, S. and Henkel, C. (2008). *Science* **320**, p. 1611.

[67] Murphy, M. T., Flambaum, V. V., Webb, J. K., Dzuba, V. V., Prochaska, J. X. and Wolfe, A. M. (2004). in S. G. Karshenboim and E. Peik (eds.), *Astrophysics, Clocks and Fundamental Constants, Lecture Notes in Physics*, Vol. 648 (Springer-Verlag, Berlin), p. 131.

[68] Murphy, M. T., Webb, J. K. and Flambaum, V. V. (2003). *MNRAS* **345**, p. 609.

[69] Murphy, M. T., Webb, J. K. and Flambaum, V. V. (2007). *Phys. Rev. Lett.* **99**, p. 239001.

[70] Murphy, M. T., Webb, J. K. and Flambaum, V. V. (2008). *MNRAS* **384**, p. 1053.

[71] Murphy, M. T., Webb, J. K. and Flambaum, V. V. (2009). *MmSAI, proc. of IAU JD 9, P. Molaro & E. Vangoni eds.* **80**.

[72] Murphy, M. T., Webb, J. K., Flambaum, V. V., Churchill, C. W. and Prochaska, J. X. (2001). *MNRAS* **327**, p. 1223.

[73] Murphy, M. T., Webb, J. K., Flambaum, V. V., Prochaska, J. X. and Wolfe, A. M. (2001). *MNRAS* **327**, p. 1237.

[74] Peik, E., Lipphardt, B., Schnatz, H., Schneider, T., Tamm, C. and Karshenboim, S. G. (2004). *Phys. Rev. Lett.* **93**, p. 170801.

[75] Petrov, Y. V., Nazarov, A. I., Onegin, M. S., Petrov, V. Y. and Sakhnovsky, E. G. (2006). *Phys. Rev. C* **74**, p. 064610.

[76] Pettini, M., Boksenberg, A., Bates, B., McCaughan, R. F. and McKeith, C. D. (1977). *A&A* **61**, p. 839.

[77] Quast, R., Reimers, D. and Levshakov, S. A. (2004). *A&A* **415**, p. L7.

[78] Reinhold, E., Buning, R., Hollenstein, U., Ivanchik, A., Petitjean, P. and Ubachs, W. (2006). *Phys. Rev. Lett.* **96**, p. 151101.

[79] Reynaud, S., Salomon, C. and Wolf, P. (2009). Testing General Relativity with Atomic Clocks, *Sp. Sci. Rev* , p. 58.

[80] Rocha, G., Trotta, R., Martins, C. J. A. P., Melchiorri, A., Avelino, P. P., Bean, R. and Viana, P. T. P. (2004). *MNRAS* **352**, p. 20.

[81] Rosenband, T., Hume, D. B., Schmidt, P. O., Chou, C. W., Brusch, A., Lorini, L., Oskay, W. H., Drullinger, R. E., Fortier, T. M., Stalnaker, J. E., Diddams, S. A., Swann, W. C., Newbury, N. R., Itano, W. M., Wineland, D. J. and Bergquist, J. C. (2008). *Science* **319**, p. 1808.

[82] Savedoff, M. P. (1956). *Nature* **178**, p. 688.

[83] Shelkovnikov, A., Butcher, R. J., Chardonnet, C. and Amy-Klein, A. (2008). *Phys. Rev. Lett.* **100**, p. 150801.

[84] Shlyakhter, A. I. (1976). *Nature* **264**, p. 340.

[85] Srianand, R., Chand, H., Petitjean, P. and Aracil, B. (2004). *Phys. Rev. Lett.* **92**, p. 121302.

[86] Srianand, R., Chand, H., Petitjean, P. and Aracil, B. (2007). *Physical Review Letters* **99**, p. 239002.

[87] Srianand, R., Petitjean, P., Ledoux, C., Ferland, G. and Shaw, G. (2005). *MNRAS* **362**, p. 549.

[88] Steinmetz, T., Wilken, T., Araujo-Hauck, C., Holzwarth, R., Hänsch, T. W., Pasquini, L., Manescau, A., D'Odorico, S., Murphy, M. T., Kentischer, T., Schmidt, W. and Udem, T. (2008). *Science* **321**, p. 1335.

[89] Suzuki, N., Tytler, D., Kirkman, D., O'Meara, J. M. and Lubin, D. (2003). *PASP* **115**, p. 1050.

[90] Thompson, R. I. (1975). *ApL* **16**, p. 3.

[91] Thompson, R. I., Bechtold, J., Black, J. H., Eisenstein, D., Fan, X., Kennicutt, R. C., Martins, C., Prochaska, J. X. and Shirley, Y. L. (2009). *ApJ* **703**, p. 1648.

[92] Tzanavaris, P., Webb, J. K., Murphy, M. T., Flambaum, V. V. and Curran, S. J. (2007). *MNRAS* **374**, p. 634.

[93] Uzan, J.-P. (2003). *Rev. Mod. Phys* **75**, p. 403.

[94] van Langevelde, H. J., van Dishoek, E. F., Sevenster, M. N. and Israel, F. P. (1995). *ApJ* **448**, p. L123.

[95] Varshalovich, D. A. and Levshakov, S. A. (1993). *JETP* **58**, p. L237.

[96] Varshalovich, D. A. and Potekhin, A. Y. (1994). *Ast. Lett.* **20**, p. 771.

[97] Verheijen, M. A. W., Oosterloo, T. A., van Cappellen, W. A., Bakker, L., Ivashina, M. V. and van der Hulst, J. M. (2008). in R. Minchin and E. Momjian (eds.), *The Evolution of Galaxies Through the Neutral Hydrogen Window, AIPC Series*, Vol. 1035, p. 265.

[98] Webb, J. K., Flambaum, V. V., Churchill, C. W., Drinkwater, M. J. and Barrow, J. D. (1999). *Phys. Rev. Lett.* **82**, p. 884.

[99] Webb, J. K., King, J. A., Murphy, M. T., Flambaum, V. V., Carswell, R. F. and Bainbridge, M. B. (2010). *arXiv:1008.3907* .

[100] Webb, J. K., Murphy, M. T., Flambaum, V. V., Dzuba, V. A., Barrow, J. D., Churchill, C. W., Prochaska, J. X. and Wolfe, A. M. (2001). *Phys. Rev. Lett.* **87**, p. 091301.

[101] Wendt, M. and Molaro, P. (2010). *arXiv:1009.3133* .

[102] Whitmore, J. B., Murphy, M. T. and Griest, K. (2010). *ApJ, submitted, (arXiv:1003.3325)* .

[103] Wiklind, T. and Combes, F. (1995). *A&A* **299**, p. 382.

[104] Wiklind, T. and Combes, F. (1996). *A&A* **315**, p. 86.

[105] Wiklind, T. and Combes, F. (1997). *A&A* **328**, p. 48.

[106] Williams, J. G., Turyshev, S. G. and Boggs, D. H. (2004). *Phys. Rev. Lett.* **93**, p. 261101.

[107] Wolfe, A. M., Broderick, J. J., Condon, J. J. and Johnston, K. J. (1976). *ApJ* **208**, p. L47.

Chapter 4

Averaging the inhomogeneous universe

Aseem Paranjape[1]

Tata Institute of Fundamental Research,
Homi Bhabha Road, Colaba, Mumbai 400005.

Abstract: A basic assumption of modern cosmology is that the universe is homogeneous and isotropic on the largest observable scales. This greatly simplifies Einstein's general relativistic field equations applied at these large scales, and allows a straightforward comparison between theoretical models and observed data. However, Einstein's equations should ideally be imposed at length scales comparable to, say, the solar system, since this is where these equations have been tested. We know that at these scales the universe is highly inhomogeneous. It is therefore essential to perform an explicit averaging of the field equations in order to apply them at large scales. It has long been known that due to the nonlinear nature of Einstein's equations, any explicit averaging scheme will necessarily lead to corrections in the equations applied at large scales.

Estimating the magnitude and behavior of these corrections is a challenging task, due to difficulties associated with defining averages in the context of general relativity (GR). It has recently become possible to estimate these effects in a rigorous manner, and we will review some of the averaging schemes that have been proposed in the literature. A tantalizing possibility explored by several authors is that the corrections due to averaging may in fact account for the apparent acceleration of the expansion of the universe. We will explore this idea, reviewing some of the work done in the literature to date. We will argue however, that this rather attractive idea is in fact *not* viable as a solution of the dark energy problem, when confronted with observational constraints.

---

[1]Current address: The Abdus Salam International Centre for Theoretical Physics, Strada Costiera 11, 34014 Trieste, Italy. E-mail: aparanja@ictp.it.

4.1   Introduction

Cosmology in the first half of the 20th century was the realm of stalwarts such as Einstein, de Sitter, Lemaître, Friedmann, Hubble and others. The simple models of a Universe described by the homogeneous and isotropic geometries characterized by the Friedmann-Lemaître-Robertson-Walker (FLRW) metric, were successful in describing the then limited amount of cosmological observations. These models were subsequently combined with the primordial nucleosynthesis models of Gamow and co-workers to form what is known as the "Big Bang" model of cosmology, which in its present form posits that the universe went through a very hot dense phase at early times and cooled as it expanded, with tiny fluctuating inhomogeneities in the past that have grown to form structures such as galaxies today [1].

A central assumption in this model of cosmology is that the universe on large scales is homogeneous and isotropic [2]. This assumption leads to tremendous simplifications in the application of general relativity (GR) to cosmology, since it reduces the ten independent components of the metric of spacetime $g_{ab}(t, \vec{x})$ to essentially a single function of time $a(t)$ known as the scale factor. In the early days of 20th century cosmology, the assumption of homogeneity and isotropy was largely motivated on grounds of simplicity and aesthetic appeal. In recent times however, it has become possible to confront this assumption with observations, which remarkably appears to be justified to a large extent (based on observations of the cosmic microwave background radiation (CMBR) [3, 4], and on analyses of galaxy surveys [5, 6, 7], although see Ref. [8]). This indicates that a model based on essentially a single function of time might in fact go a long way in furthering our understanding of the behavior of the universe.

Of course the real universe is not homogeneous; we see a rich variety of structure around us from stellar systems to galaxies to clusters of galaxies and even larger structures [9]. Perhaps one of the biggest successes of cosmological theory based on GR, has been the explanation of how statistical properties of the large scale structure arise [10]. The relevant calculations are largely based on linear perturbation theory (i.e. linearizing Einstein's equations around the smooth FLRW solution) which is valid at all length scales of interest at early times and on large scales at late times [11, 12]. Dynamics on small scales at late times involves nonlinear theory, and is dealt with using approximation schemes such as the Press-Schechter formalism and its extensions [13], "Newtonian" nonlinear perturbation analyses [14] and numerical simulations [15]. While such treatments have met with great success in the description of the statistical properties of the anisotropies in the temperature of the CMBR, as well as of the inhomogeneous distribution of galaxies, there are two causes for concern.

The first is a purely theoretical issue, and will occupy us for the rest of this chapter. The idea that the large scale universe is homogeneous and isotropic necessarily entails an implicit notion of averaging on these large scales. In other words, what one is really saying is, "When the spatially fluctuating parts of the solution of GR describing our universe are averaged out, what is left is the homogeneous and isotropic FLRW solution of Einstein's equations". The immediately obvious problem with this statement is that the details of the averaging operation are not at all clear, and indeed are usually never specified. A bigger problem is one noted by Ellis [16], and can be stated in the following symbolic way. If g denotes the metric, Γ the Christoffel connection and $E[g]$ the Einstein tensor for the metric g, then we have the relations

$$\Gamma \sim \partial g \; ; \; E[g] \sim \partial \Gamma + \Gamma^2 \,, \tag{4.1}$$

with ∂ denoting spacetime derivatives. The Einstein equations are therefore

$$E[g] = T \,, \tag{4.2}$$

with T denoting the energy-momentum tensor of the matter components. Now, irrespective of any details of the averaging operation, one notes that

$$E[\langle g \rangle] - \langle E[g] \rangle \sim \langle \Gamma \rangle^2 - \langle \Gamma^2 \rangle \neq 0 \,, \tag{4.3}$$

with the angular brackets denoting the averaging. The FLRW solution would amount to solving the equations $E[\langle g \rangle] = \langle T \rangle$. In general therefore, it is *not* true that averaging out the fluctuating inhomogeneities leaves behind the FLRW solution, since what we are actually left with is

$$E[\langle g \rangle] = \langle T \rangle - \mathcal{C} \; ; \; \mathcal{C} \sim \langle \Gamma^2 \rangle - \langle \Gamma \rangle^2 \,, \tag{4.4}$$

and the homogeneous solution that we are looking for will depend on the details of the correction terms $\mathcal{C}$.

The second cause for concern comes from observations. It has now been established beyond a reasonable doubt, that the FLRW metric confronted with observations indicates an accelerating scale factor [17]. Conventional sources of energy such as radiation and nonrelativistic matter cannot explain the acceleration, and it is now common to attribute this effect to a hitherto unknown "dark energy", which in its simplest form is a cosmological constant. The true nature of this additional component in the cosmological equations, is perhaps the most challenging puzzle facing both theorists and observers today. A huge amount of research has gone into (a) explaining the value that a cosmological constant term must take to explain data or (b) assuming a zero cosmological constant, constructing models of a dynamical dark energy which explains the observed acceleration [18]. It is fair to say however that there is no theoretical consensus on what the origin of dark energy is. Since we have seen above that the effects of averaging lead to some extra, as yet unknown terms in the equations, it is natural to ask whether these two issues are connected. Could the acceleration of the universe be explained by the effects of averaging inhomogeneities ("backreaction") in the universe? Regardless of the answer to this question, what is the nature and magnitude of this backreaction?

Estimating the magnitude and behavior of these corrections is a challenging task, due to difficulties associated with defining averages in the context of general relativity (GR). It has recently become possible to estimate these effects in a rigorous manner, and we will review some of the averaging schemes that have been proposed in the literature. A comprehensive treatment of inhomogeneous solutions of GR, as well as several references and a discussion on the averaging problem can be found in the book by Krasiński [19]. Throughout, we will set the speed of light to unity, and use lowercase Latin indices $a, b, ..i, j, ..$ for spacetime indices and uppercase Latin indices $A, B, ..I, J, ..$ for spatial indices.

4.2   History of the averaging problem

The problem of averaging in general relativity has a history going back even further than Ellis' work of 1984. In the context of gravitational radiation, the problem of second order effects of gravity waves on the large scale background metric of spacetime was studied by Isaacson in the 1960's [20] in the "short-wavelength" approximation. Isaacson used an averaging operation which he called the "BH assumption" after Brill and Hartle [21], which was suited to studying the effects of perturbative gravity waves in a spacetime region encompassing many wavelengths. An attempt to generalize Isaacson's results was made by Noonan [22], who introduced a different averaging procedure which was also constructed for situations where inhomogeneities were perturbative in nature, the goal being to define a consistent gravitational energy-momentum pseudotensor in the presence of matter.

4.2.1 *Noonan's averaging scheme*

Noonan's basic premise was the postulate that there exist two types of observers, a microscopic observer O' and a macroscopic observer O. To each observer there corresponds a metric (g'_{ab} for the microscopic, g_{ab} for the macroscopic) and a matter energy-momentum tensor (T'_{ab} and T_{ab}). The tensors for the macroscopic observer were assumed to vary on scales larger than the corresponding scales for the microscopic observer. By excluding those coordinate transformations which do not change the macroscopic tensors appreciably while significantly affecting the microscopic tensors, the *difference* between the microscopic and macroscopic objects would also behave as tensors under the remaining transformations. Noonan then defined the following averaging operation for any microscopic object Q',

$$\langle Q' \rangle = \frac{\int d^4x \sqrt{-g} Q'}{\int d^4x \sqrt{-g}} \approx \frac{\int d^4x Q'}{\int d^4x} \tag{4.5}$$

where the second approximate equality follows since the macroscopic metric g_{ab} (and hence its determinant g) does not change appreciably over the averaging domain which is small compared to macroscopic scales but large compared to microscopic scales. This averaging was applied to an expansion of the Einstein equations in powers of the difference $h_{ab} \equiv g'_{ab} - g_{ab}$, and a formula was derived for the gravitational energy-momentum pseudotensor in the presence of matter. The averaging operation in Eqn. (4.5) sidesteps an issue which should in fact be handled more carefully. Namely, when one is trying to develop an averaging scheme for tensors which will also be used to average the metric, one needs to deal with the tricky issue of how to self consistently define an invariant volume element (e.g., $\sqrt{-g} d^4x$ above) which will be used in the averaging integrals. Noonan's averaging volume element uses the *averaged* or macroscopic metric, and hence runs into a potential circularity problem when one wants to average the microscopic metric itself (the averaging operation above is defined by a quantity which itself can only be defined after an averaging is performed). As seen in the second approximate inequality in Eqn. (4.5), this problem was circumvented by removing all reference to the averaged metric, and self consistency imposed by requiring that the macroscopic metric g_{ab} be given by

$$g_{ab} = \langle g'_{ab} \rangle = \frac{\int d^4x g'_{ab}}{\int d^4x} . \tag{4.6}$$

As a consequence of assuming this approximate definition of the averaging operation, one is forced to consider only those microscopic spacetimes which are perturbations about the macroscopic spacetime. This work, while not intended for cosmological purposes, nevertheless demonstrates the potential problems that arise in defining an averaging operation in which the background geometry itself has to be averaged.

4.2.2   *Futamase's scheme*

Interest in the cosmological consequences of such an averaging picked up only after Ellis very clearly laid down the problems and possibilities that open up when the idea of averaging in general relativity is taken seriously. An example is the work of Futamase [23, 24], who introduced a spatial averaging procedure after performing a $3 + 1$ splitting of spacetime, and computed backreaction terms arising from averaging second order perturbations, finding them to be negligibly small. In Ref. [23] the metric of spacetime was assumed to be of the form $g_{ab} = g_{ab}^{(\mathrm{FLRW})} + h_{ab}$, where $g_{ab}^{(\mathrm{FLRW})}$ is the FLRW metric and h_{ab} is a small perturbation. A spatial averaging was defined using the proper spatial 3-volume element of the background FLRW spacetime. This averaging was then used to calculate contributions to the evolution equations of the background scale factor, from averages of quantities quadratic in h_{ab}. The basic idea was that nonlinearities of the matter distribution would affect the local expansion rate of the inhomogeneously perturbed metric, and these effects would be expected to show up in the averages of the second order quantities. In later work, Futamase [24] used a more sophisticated averaging scheme, using ideas from Isaacson's earlier work, in the context of a $3 + 1$ splitting of spacetime, and studied a perturbed FLRW metric in this framework. The equations derived in Ref. [24] coincided with those derived by Futamase in earlier work, with the added feature of being invariant under transformations which left the background FLRW spacetime fixed.

4.2.3   *Boersma's scheme*

Another example is the work of Boersma [25], who attempted to construct a gauge-invariant (i.e. coordinate independent) averaging procedure in perturbation theory, and also estimated that backreaction effects remain negligibly small at the present epoch. This work dealt explicitly with the issue of a choice of averaging operation, and appropriate choices of gauge, within the perturbed FLRW framework. Boersma chose a linearized averaging operation for the perturbation h_{ij} around an FLRW metric $g_{ij}^{(\mathrm{FLRW})}$

by requiring that the homogeneous and isotropic FLRW metric be a stable fixed point of the averaging operation. This averaging operation was shown to reduce to a spatial averaging of the scalar quantities h_{00} and h_A^A in a background which was *synchronous* ($g_{00}^{(\text{FLRW})} = -1$). The modified Friedmann equation was derived after gauge fixing the perturbation variables and their averages, and the corrections were related to observational quantities determinable using matter power spectrum data. Boersma showed that the corrections to the standard Friedmann equation in this approach were $\sim 10^{-5}\text{-}10^{-4}$ smaller than the zero order contributions.

4.2.4 *Kasai's scheme*

An early attempt at treating general inhomogeneous spacetimes which would behave like FLRW on average (without assuming the inhomogeneities to be small) was by Kasai [26]. The spatial averaging used here was quite general and did not assume deviations from FLRW behavior in the inhomogeneous metric to be small (although it was strictly consistent only for scalar quantities). However, the author assumed certain correlation terms to vanish, these terms being of the form $\Theta_B^A \Theta_A^B - \Theta^2$, where $(-\Theta_B^A)$ is the extrinsic curvature of a spatial hypersurface in a given $3 + 1$ splitting (see Buchert's averaging scheme below), and $\Theta \equiv \Theta_A^A$. As we will see, it is precisely such terms which are expected to lead to nontrivial effects of averaging in the effective Einstein equations, and not surprisingly, Kasai concluded that this particular model of inhomogeneities behaves on average exactly like the standard FLRW model with nonrelativistic matter. (For other work on the averaging problem, see Ref. [27].)

4.2.5 *Conventional wisdom and controversy*

It may seem intuitively obvious that perturbatively small inhomogeneities can only lead to negligibly small backreaction effects. Indeed, this has been the conventional wisdom on this subject, and has recently been spelt out by Ishibashi and Wald [28]. One starts by assuming that inhomogeneities in the universe can be described in the Newtonian approximation of GR, by the gravitational potential $\varphi(t, \vec{x})$ with $|\varphi| \ll 1$, which appears in the perturbed FLRW metric

$$ds^2 = -(1 + 2\varphi)d\tau^2 + a(\tau)^2(1 - 2\varphi)d\vec{x}^2 \,, \tag{4.7}$$

and satisfies the Poisson equation $\nabla^2\varphi = 4\pi G a^2 \delta\rho$ where $\delta\rho(t, \vec{x})$ is the fluctuation of matter density about the mean homogeneous value $\bar{\rho}(t)$, and can in general have a large value. (E.g., in clusters of galaxies one finds $\delta \equiv \delta\rho/\bar{\rho} \sim 10^2$, and the ratio increases on smaller length scales). One then argues that the universe we observe *does* seem to be very well-described by the above model, and effects of averaging this model can only arise at second order in φ and should hence be extremely small.

There is a loophole in this argument though. The catch is that the background expansion $a(t)$ is defined completely ignoring the backreaction, which is an integrated effect with contributions from a large range of length scales. This means that the following possibility cannot be *a priori* ruled out : Initial conditions are specified as a perturbation around a specific FLRW solution, but the integrated effect of the backreaction grows (with time) in such a manner as to effectively yield a late time solution which is a perturbation around a *different* FLRW model. Indeed, there are calculations in the literature that do indicate that this may happen. For example, Martineau and Brandenberger [29] showed in a toy model that long wavelength fluctuations can give rise to a backreaction contribution which has a late-time effective equation of state similar to a cosmological constant. Their calculations were based on the averaging procedure developed by Abramo et al. [30] in the context of backreaction in inflationary cosmology. Other claims to solving the dark energy problem using backreaction from long wavelength fluctuations were made by Barausse et al. [31] and Kolb et al. [32]. It is fair to say however, that such claims have been controversial. A number of authors have argued that when effects of long wavelength fluctuations are suitably "renormalized" and the background suitably redefined, the backreaction cannot lead to acceleration of the scale factor [33]. Nevertheless, what is definitely true is that the idea of backreaction of cosmological fluctuations has generated a lively debate in the community [28, 34, 35, 36].

This article will not deal with the effects of long wavelength fluctuations, although we will see that certain assumptions need to be made in order to define a self-consistent perturbation theory in the presence of an averaging operation. A separate and equally interesting question, which will be our main focus, is whether cosmological perturbation theory is *stable* in the presence of the backreaction contribution. There are results in the literature which indicate that this might not be the case, and that the backreaction can grow with time in such a manner that at late times (when matter fluctuations have become nonlinear) perturbation theory *in the met-*

ric also no longer holds [35] (see also Ref. [37]). In the same vein, there are arguments using *nonperturbative* toy models of gravitational collapse and nonlinear structure formation, which suggest that perturbation theory may not give correct insight into gravitational dynamics at late times in cosmology [38]. If these results are relevant for the real world, then it not only means that the conventional wisdom is badly failing, but in fact implies that all of late-time cosmology must be reworked from scratch (see, e.g. Ref. [39]; also see however Ref. [40]). On the other hand, if these results are for some reason or other not realistic, then it is important to ask what is wrong with such arguments, and further what the correct approach to the problem is.

Clearly, to make any headway in this problem, it is first essential to have a reliable averaging scheme at hand. Since the questions one is asking deal with the stability of cosmological perturbation theory, this averaging scheme needs to be inherently *nonperturbative*, i.e. the validity of perturbation theory should not be a prerequisite to defining the averaging prescription. We will deal with two averaging schemes present in the literature : the spatial averaging of scalars defined by Buchert [41, 42, 43], and the fully covariant tensor averaging defined by Zalaletdinov [44, 45, 46]. Some very interesting early work on possible nonperturbative effects of averaging was by Buchert and Ehlers [47], followed up by Ref. [48], in the context of spatial averaging in Newtonian cosmology. Buchert's averaging operation in general relativity has since been used by several authors to explore the effects of backreaction in various situations [39, 49, 50, 51, 52], including the perturbative contexts mentioned above [35, 37], and has also been compared against observations [53]. As we shall see later, this averaging scheme has an appealing simplicity of implementation, which could be a reason for the amount of attention it has received. In contrast, Zalaletdinov's averaging scheme (which was developed earlier than Buchert's work) is technically rather challenging to handle and involves a fair amount of complicated algebra. Its strength however lies in the fact that it is a fully covariant prescription which, at the end of the day, yields an object which can be legitimately called the "averaged metric" on an "averaged manifold". This ultimately allows us to make physically clear statements regarding the backreaction, which is difficult to do in Buchert's scheme as it stands. While Zalaletdinov's scheme has not received the same amount of attention as Buchert's, there *has* been a series of very interesting results derived in this framework by Zalaletdinov and coworkers [54]. In order to keep this article concise, we will describe some of the mathematical details of Buchert's scheme, which is easier to follow, and only mention some of the salient features of Zalaletdinov's scheme. Once the basic structure of the backreaction is in place, we will present results based on calculations in Zalaletdinov's scheme, referring the reader to the literature for details.

4.3 Buchert's spatial averaging of scalars

The most straightforward and intuitively clear application of Buchert's spatial averaging is in the case when the matter source is a pressureless "dust" with an energy-momentum tensor $T^{ab} = \rho u^a u^b$, with u^a the dust 4-velocity which satisfies $u_a u^a = -1$. Assuming further that the dust is irrotational, the 4-velocity will be orthogonal to 3-dimensional spatial sections and the metric can be written in "synchronous and comoving" coordinates (in which $u^a = (1, \vec{0})$) [55] as,

$$ds^2 = -dt^2 + h_{AB}(t, \vec{x}) dx^A dx^B . \tag{4.8}$$

The expansion tensor Θ^A_B is given by $\Theta^A_B \equiv (1/2) h^{AC} \dot{h}_{CB}$ where the dot refers to a derivative with respect to time t. The traceless symmetric shear tensor is defined as $\sigma^A_B \equiv \Theta^A_B - (\Theta/3)\delta^A_B$ where $\Theta = \Theta^A_A$ is the expansion scalar. The Einstein equations can be split into a set of scalar equations and a set of vector and traceless tensor equations. The scalar equations are the Hamiltonian constraint (4.9a) and the evolution equation for Θ (4.9b),

$$^{(3)}\mathcal{R} + \frac{2}{3}\Theta^2 - 2\sigma^2 = 16\pi G\rho , \tag{4.9a}$$

$$^{(3)}\mathcal{R} + \dot{\Theta} + \Theta^2 = 12\pi G\rho , \tag{4.9b}$$

where $^{(3)}\mathcal{R}$ is the Ricci scalar of the 3-dimensional hypersurface of constant t and σ^2 is the rate of shear defined by $\sigma^2 \equiv (1/2)\sigma^A_B \sigma^B_A$. Eqns. (4.9a) and (4.9b) can be combined to give Raychaudhuri's equation

$$\dot{\Theta} + \frac{1}{3}\Theta^2 + 2\sigma^2 + 4\pi G\rho = 0 . \tag{4.10}$$

The continuity equation $\dot{\rho} = -\Theta\rho$ which gives the evolution of ρ, is consistent with Eqns. (4.9a), (4.9b). We only consider the scalar equations, since the spatial average of a scalar quantity can be defined in a gauge covariant manner within a given foliation of spacetime. For the spacetime described by (4.8), the spatial average of a scalar $\Psi(t, \vec{x})$ over a *comoving* domain $\mathcal{D}$ at time t is defined by

$$\langle \Psi \rangle_{\mathcal{D}} = \frac{1}{V_{\mathcal{D}}} \int_{\mathcal{D}} d^3x \sqrt{h}\, \Psi , \tag{4.11}$$

where h is the determinant of the 3-metric h_{AB} and V_D is the volume of the comoving domain given by $V_D = \int_D d^3x \sqrt{h}$. The following commutation relation then holds [41]

$$\langle \Psi \rangle_D^{\cdot} - \langle \dot{\Psi} \rangle_D = \langle \Psi \Theta \rangle_D - \langle \Psi \rangle_D \langle \Theta \rangle_D, \qquad (4.12)$$

which yields for the expansion scalar Θ

$$\langle \Theta \rangle_D^{\cdot} - \langle \dot{\Theta} \rangle_D = \langle \Theta^2 \rangle_D - \langle \Theta \rangle_D^2. \qquad (4.13)$$

Introducing the dimensionless scale factor $a_D \equiv (V_D/V_{Din})^{1/3}$ normalized by the volume of the domain D at some initial time t_{in}, we can average the scalar Einstein equations (4.9a), (4.9b) and the continuity equation to obtain [41]

$$\left(\frac{\dot{a}_D}{a_D}\right)^2 = \frac{8\pi G}{3}\langle \rho \rangle_D - \frac{1}{6}\left(Q_D + \langle R \rangle_D\right), \qquad (4.14a)$$

$$\left(\frac{\ddot{a}_D}{a_D}\right) = -\frac{4\pi G}{3}\langle \rho \rangle_D + \frac{1}{3}Q_D, \qquad (4.14b)$$

$$\langle \rho \rangle_D^{\cdot} = -\langle \Theta \rangle_D \langle \rho \rangle_D = -3\frac{\dot{a}_D}{a_D}\langle \rho \rangle_D. \qquad (4.14c)$$

Here $\langle R \rangle_D$, the average of the spatial Ricci scalar $^{(3)}R$, is a domain dependent spatial constant. The 'backreaction' Q_D is given by

$$Q_D \equiv \frac{2}{3}\left(\langle \Theta^2 \rangle_D - \langle \Theta \rangle_D^2\right) - 2\langle \sigma^2 \rangle_D, \qquad (4.15)$$

and is also a spatial constant. The last equation (4.14c) simply reflects the fact that the mass contained in a comoving domain is constant by construction : since $\Theta = \partial_t \ln \sqrt{h}$, the local continuity equation $\dot{\rho} = -\Theta \rho$ can be solved to give $\rho \sqrt{h} = \rho_0 \sqrt{h_0}$ where the subscript 0 refers to some arbitrary reference time t_0. The mass M_D contained in a comoving domain D is then $M_D = \int_D \rho \sqrt{h} d^3x = \int_D \rho_0 \sqrt{h_0} d^3x = $ constant. Hence

$$\langle \rho \rangle_D = \frac{M_D}{V_{Din} a_D^3} \qquad (4.16)$$

which is precisely what is implied by Eqn. (4.14c). Equations (4.14a), (4.14b) can be compared with the Friedmann equations

$$\left(\frac{1}{a}\frac{da}{d\tau}\right)^2 = \frac{8\pi G}{3}\rho_{\text{FLRW}} - \frac{k}{a^2} \ ; \ \frac{1}{a}\frac{d^2a}{d\tau^2} = -\frac{4\pi G}{3}(\rho_{\text{FLRW}} + 3p_{\text{FLRW}}), \quad (4.17)$$

where $a(\tau)$ and k appear in the unperturbed FLRW metric

$$ds^2 = -d\tau^2 + a(\tau)^2\left(\frac{dr^2}{1-kr^2} + r^2 d\Omega^2\right), \qquad (4.18)$$

and ρ_{FLRW} and p_{FLRW} are the homogeneous FLRW energy density and pressure respectively. Equations (4.14a), (4.14b) can then be thought of as "modified Friedmann equations", with the modifications arising due to the presence of the backreaction $\mathcal{Q}_{\mathcal{D}}$ and the fact that the averaged Ricci curvature in general need not evolve like $\sim a_{\mathcal{D}}^{-2}$ as in the FLRW case.

A necessary condition for (4.14b) to integrate to (4.14a) takes the form of the following differential equation involving $\mathcal{Q}_{\mathcal{D}}$ and $\langle \mathcal{R} \rangle_{\mathcal{D}}$

$$\dot{\mathcal{Q}}_{\mathcal{D}} + 6\frac{\dot{a}_{\mathcal{D}}}{a_{\mathcal{D}}}\mathcal{Q}_{\mathcal{D}} + \langle \mathcal{R} \rangle_{\mathcal{D}}^{\cdot} + 2\frac{\dot{a}_{\mathcal{D}}}{a_{\mathcal{D}}}\langle \mathcal{R} \rangle_{\mathcal{D}} = 0, \qquad (4.19)$$

which is a very interesting equation because it shows that the evolution of the backreaction is intimately tied to that of the average spatial curvature. Scaling solutions for this equation have been explored by Buchert, Larena and Alimi [51], a simple example being $\langle \mathcal{R} \rangle_{\mathcal{D}} \propto a_{\mathcal{D}}^{-2}$, $\mathcal{Q}_{\mathcal{D}} \propto a_{\mathcal{D}}^{-6}$. Clearly the FLRW solution with $\mathcal{Q}_{\mathcal{D}} = 0$ is a special case. The criterion to be met in order for the effective scale factor $a_{\mathcal{D}}$ to accelerate, is

$$\mathcal{Q}_{\mathcal{D}} > 4\pi G \langle \rho \rangle_{\mathcal{D}}. \qquad (4.20)$$

Equations (4.14) and (4.19) describe the essence of Buchert's averaging formalism, for the simplest case of irrotational dust. These equations have been analyzed by many authors. We will not go into details of calculations based on these equations in this article, since ultimately we are interested in results based on Zalaletdinov's formalism. For some examples of the kind of calculations performed using Buchert's equations, and for further references, see Ref. [56].

4.4 Zalaletdinov's Macroscopic Gravity (MG)

We now turn to Zalaletdinov's fully covariant averaging formalism, which when applied to Einstein's equations, leads to what Zalaletdinov calls the theory of Macroscopic Gravity (MG). After describing the core mathematics which defines the averaging operation, we will argue that in the cosmological context one must necessarily deal with a *spatial averaging limit* of this averaging. The reader is referred to the original literature as cited throughout this section for details of the calculations.

Zalaletdinov [44] tackled the averaging problem headlong by considering the most general question – given a four dimensional spacetime manifold $\mathcal{M}$ with a metric g_{ab}, how does one construct and determine the properties of an *averaged manifold* $\bar{\mathcal{M}}$? Zalaletdinov solved this problem by first

defining a complete *bilocal exterior calculus* on the base manifold $\mathcal{M}$.[2].
Here we will only mention some of the main results derived by Zalaletdinov
and co-workers.

A bilocal operator $\mathcal{W}_j^{a'}(x', x)$ and its inverse $\mathcal{W}_{a'}^j(x, x')$ satisfying certain properties were defined and shown to exist [45] on general differentiable
manifolds. Hereafter, the primed index refers to the point x' and the un-
primed index to the point x, and the bivector $\mathcal{W}_j^{a'}(x', x)$ transforms like
a vector at x' and a co-vector at x. This bivector serves the following
purposes in MG –

- $\mathcal{W}_j^{a'}$ is used to define the bilocal extension of a tensorial object.
  This is best seen with the help of an example. If $P_b^a(x)$ is a $(1, 1)$
  tensor on $\mathcal{M}$, then its bilocal extension $\widetilde{P}_b^a$ is defined as

$$\widetilde{P}_b^a(x', x) = \mathcal{W}_b^{b'}(x', x)\mathcal{W}_{a'}^a(x, x')P_{b'}^{a'}(x') \,. \qquad (4.21)$$

  Using the definition of the bilocal extension, the average of say P_b^a
  over a spacetime region Σ with a supporting point x, is then given
  by

$$\bar{P}_b^a(x) = \langle \widetilde{P}_b^a \rangle = \frac{1}{V_\Sigma} \int_\Sigma d^4x' \sqrt{-g'} \widetilde{P}_b^a(x', x) \ ; \ V_\Sigma = \int_\Sigma d^4x' \sqrt{-g'} \,,$$
$$(4.22)$$

  The conditions imposed on the bivector $\mathcal{W}_j^{a'}$ ensure that the average of any tensor field on $\mathcal{M}$ is itself a smooth tensor field on
  $\mathcal{M}$.

- Secondly, $\mathcal{W}_j^{a'}$ is used to specify a *Lie dragging* of the averaging
  region Σ. This ensures that the volumes of the averaging regions
  constructed at nearby supporting points are coordinated in a well
  defined manner, and completely defines the exterior derivatives of
  averaged quantities (see Ref. [44]). Suppose x^a and $x^a + \xi^a \Delta\lambda$ are
  the coordinates of two support points, where $\Delta\lambda$ is a small change
  in the parameter along the integral curve of a given vector field ξ^a.
  Symbolically denote the two points as x and $x + \xi\Delta\lambda$. Then the
  averaging region at $x + \xi\Delta\lambda$ is defined in terms of the averaging
  region Σ at x, by transporting every point $x' \in \Sigma$ around x along
  the appropriate integral curve of a *new* bilocal vector field $S^{a'}$
  defined as $S^{a'}(x', x) = \mathcal{W}_j^{a'}(x', x)\xi^j(x)$, thereby constructing the

---

[2]For a very nice pictorial treatment of differential forms and exterior calculus as applied
in standard GR, see the book by Misner, Thorne and Wheeler [57]. A *bilocal* exterior
calculus requires the differential forms being considered to be functions of *two* points x
and x' of the manifold $\mathcal{M}$.

averaging region $\Sigma(\Delta\lambda)$ with support point $x + \xi\Delta\lambda$. (See Refs. [44, 45] for further discussion on the significance of this averaging region coordination.)

The explicit form of the "coordination bivector" $\mathcal{W}_j^{a'}(x', x)$ is given by

$$\mathcal{W}_j^{a'}(x', x) = f_m^{a'}(x') f^{-1}{}_j^m(x), \tag{4.23}$$

with

$$f_m^a(x(\phi^n)) = \frac{\partial x^a}{\partial\phi^m} \quad ; \quad f^{-1}{}_j^m(\phi(x^k)) = \frac{\partial\phi^m}{\partial x^j}, \tag{4.24}$$

where x^a are the coordinates we choose to work in, and ϕ^m are a set of "volume preserving coordinates" in which the metric determinant is constant $g(\phi^m) = $ constant. When expressed in terms of such a volume preserving coordinate (VPC) system, the coordination bivector takes its most simple form, namely

$$\mathcal{W}_j^{a'}(x', x)\,|_{\text{proper}} = \delta_j^{a'}. \tag{4.25}$$

The averaging operation so defined is then used to *construct* an "averaged" differentiable manifold $\bar{\mathcal{M}}$. This is basically done by showing that the average of the affine connection on $\mathcal{M}$ itself behaves like a connection, and is therefore postulated as the affine connection for the abstract manifold $\bar{\mathcal{M}}$. One defines the (tensorial) connection correlation terms,

$$Z^a{}_{b[m}{}^i{}_{\underline{j}n]} = \langle \tilde{\Gamma}^a_{b[m} \tilde{\Gamma}^i_{\underline{j}n]} \rangle - \langle \tilde{\Gamma}^a_{b[m} \rangle \langle \tilde{\Gamma}^i_{\underline{j}n]} \rangle \quad ; \quad Z^a{}_{ijb} = 2Z^a{}_{ik}{}^k{}_{jb}, \tag{4.26}$$

where the square brackets denote antisymmetrization and the underlined indices are not antisymmetrized. After some tedious algebra and a few simplifying assumptions, it can then be shown that the averaged Einstein equations (with the sign convention used in Ref. [44]) read

$$E_b^a = -\kappa T_b^a + C_b^a, \tag{4.27}$$

where E_b^a is the Einstein tensor for $\bar{\mathcal{M}}$, T_b^a is the averaged energy-momentum tensor and the correlation tensor C_b^a is defined as

$$C_b^a = \left(Z^a{}_{ijb} - \frac{1}{2}\delta_b^a Z^m{}_{ijm} \right) G^{ij}, \tag{4.28}$$

where $G_{ab} = \langle g_{ab} \rangle$ is the metric on $\bar{\mathcal{M}}$ (it is shown in Ref. [44] that this choice can always be made) and G^{ab} is its inverse.

One simplifying assumption that has gone into writing Eqns. (4.27) and (4.28) is that the averaged manifold $\bar{\mathcal{M}}$ is a highly symmetric space (in the cosmological context, e.g., one would assume $\bar{\mathcal{M}}$ to be the FLRW spacetime). This ensures that the inverse of the averaged metric G_{ab} is the same as the average of the inhomogeneous inverse metric, namely $G^{ab} = \langle \widetilde{g}^{ab} \rangle$, which would not be true in general.

Another assumption made is that the correlation terms defined in Eqn. (4.26) are the only ones that contribute to the modified equations. The general MG formalism also accounts for some additional correction terms necessary to close the set of differential equations in the theory. These terms (which arise as a differentiable 3-form and a 4-form, compared with the 2-form terms of Eqn. (4.26)) complicate the formalism considerably. Fortunately though, setting these additional terms to zero is a consistent assumption.

The two assumptions described in the preceding paragraphs are not essential to the formalism, they only make life a little easier. There is, however, an assumption related to the averages of the metric and the connection on $\mathcal{M}$, which is crucially used in writing the equations of MG even in the most general case. This basically states that whenever one averages the product of the connection with a "slowly varying" tensor (defined to be either a covariantly constant tensor such as the metric, or a Killing tensor, etc.), one can effectively "pull out" the slowly varying object from the average. To be more precise, if $c_{b\ldots}^{a\ldots}(x)$ is the slowly varying object, then one assumes

$$\langle \widetilde{c}_{b\ldots}^{a\ldots}\widetilde{\Gamma}_{jk}^{i} \rangle = \langle \widetilde{c}_{b\ldots}^{a\ldots} \rangle \langle \widetilde{\Gamma}_{jk}^{i} \rangle \quad ; \quad \langle \widetilde{c}_{b\ldots}^{a\ldots}\widetilde{\Gamma}_{jk}^{i}\widetilde{\Gamma}_{mn}^{l} \rangle = \langle \widetilde{c}_{b\ldots}^{a\ldots} \rangle \langle \widetilde{\Gamma}_{jk}^{i}\widetilde{\Gamma}_{mn}^{l} \rangle . \quad (4.29)$$

This assumption would be reasonable if, e.g., one assumes that the "slowly varying" objects vary on two scales L_1 and L_2 such that $L_1 \ll L \ll L_2$ where L is the scale of the averaging domain. This assumption is also remarkable in that it is the single assumption required to average out the Bianchi identities and the Einstein equations [44].

Eqns. (4.26)-(4.28) summarize the entire MG formalism as relevant for cosmology. We henceforth assume that the averaged manifold $\bar{\mathcal{M}}$ corresponds to the FLRW spacetime, which we assume to have flat spatial sections for simplicity. One can then choose coordinates (t, x^A), $A = 1, 2, 3$, on $\bar{\mathcal{M}}$ such that the line element takes the form

$$^{(\bar{\mathcal{M}})}ds^2 = -f(t)^2 dt^2 + a^2(t)\delta_{AB}dx^A dx^B , \quad (4.30)$$

where $\delta_{AB} = 1$ for $A = B$, and 0 otherwise, and we allow $f(t)$ to be an unspecified function of time for now.

4.4.1   *A spatial averaging limit*

An important ingredient that remains to be added to the MG formalism however, is a *spatial averaging limit*. The simplest way to understand the need for such a limit, is to note that the homogeneous and isotropic FLRW spacetime *must* be left invariant under the averaging operation, and this is only possible if the averaging is tuned to the uniquely defined spatial slices of constant curvature in the FLRW spacetime. This leads us to the crucial question of the choice of *gauge* for the underlying geometry : namely, what choice of spatial sections for the *inhomogeneous* geometry, will lead to the spatial sections of the FLRW metric in the comoving coordinates defined in Eqn. (4.30)? Since the matter distribution at scale L_{inhom} need not be pressure-free (or, indeed, even of the perfect fluid form), there is clearly no natural choice of gauge available, although locally, a synchronous reference can always be chosen. We note that there must be *at least one* choice of gauge in which the averaged metric has spatial sections in the form (4.30) – this is simply a refinement of the Cosmological Principle, and of the Weyl postulate, according to which the universe is homogeneous and isotropic on large scales, and individual galaxies are considered as the "observers" travelling on trajectories with tangent ∂_τ. In the averaging approach, it makes more sense to replace "individual galaxies" with the *averaging domains* considered as physically infinitesimal cells – the "points" of the averaged manifold $\bar{\mathcal{M}}$. This is physically reasonable since we know after all, that individual galaxies exhibit peculiar motions, undergo mergers and so on. This idea is also more in keeping with the notion that the universe is homogeneous and isotropic *only on the largest scales*, which are much larger than the scale of individual galaxies.

Consider any $3 + 1$ spacetime splitting in the form of a lapse function $N(t, x^J)$, a shift vector $N^A(t, x^J)$, and a metric for the 3-geometry $h_{AB}(t, x^J)$, so that the line element on $\mathcal{M}$ can be written as

$$^{(\mathcal{M})}ds^2 = - \left(N^2 - N_A N^A \right) dt^2 + 2N_B dx^B dt + h_{AB} dx^A dx^B , \qquad (4.31)$$

where $N_A = h_{AB} N^B$. At first sight, it might seem reasonable to leave the choice of gauge arbitrary. However, it turns out that if we make the assumption that the spatial sections on $\mathcal{M}$ leading to the metric (4.30) on $\bar{\mathcal{M}}$, are spatial sections *in a volume preserving gauge*, with $N\sqrt{h} = 1$, then the correlation terms simplify greatly. This is not surprising since the MG formalism is nicely adapted to the choice of volume preserving coordinates. Moreover, as we will see later, at least in the perturbative context a modified version of this "VP gauge assumption" is in fact necessary in order to consistently set up the formalism. Note that in this gauge, the 4-dimensional average takes on a particularly simple form : for a tensor $p^i_j(x)$, with a spacetime averaging domain given by the "cuboid" Σ defined by

$$\Sigma = \left\{ (t', x^{A'}) \mid t - T/2 < t' < t + T/2, x^A - L/2 \right.$$

$$\left. < x^{A'} < x^A + L/2; A = 1, 2, 3 \right\}, \qquad (4.32)$$

where T and L are averaging time and length scales respectively, the average is given by

$$\langle \tilde{p}^i_j \rangle_{ST}(t, \mathbf{x}) = \langle p^i_j \rangle_{ST}(t, \mathbf{x})$$

$$= \frac{1}{TL^3} \int_{t-T/2}^{t+T/2} dt' \int_{x-L/2}^{x+L/2} dx' \int_{y-L/2}^{y+L/2} dy'$$

$$\times \int_{z-L/2}^{z+L/2} dz' \left[p^i_j(t', x', y', z') \right]. \qquad (4.33)$$

We define the spatial averaging limit as the limit $T \to 0$ (or $T \ll L_{\text{Hubble}}$) which is interpreted as providing a definition of the average on a spatial domain corresponding to a "thin" time slice, the averaging operation now being given by

$$\langle p^i_j \rangle(t, \mathbf{x}) = \frac{1}{L^3} \int_{x-L/2}^{x+L/2} dx' \int_{y-L/2}^{y+L/2} dy' \int_{z-L/2}^{z+L/2} dz' \left[p^i_j(t, x', y', z') \right]$$

$$+ \mathcal{O}\left(TL_{\text{Hubble}}^{-1} \right). \qquad (4.34)$$

(Note the time dependence of the integrand.)

Calculating the correlation objects is now a straightforward but tedious job. The final results can be summarized as the following modified Friedmann equations

$$H^2_{\text{FLRW}} \equiv \left(\frac{1}{a} \frac{da}{d\tau} \right)^2 = \frac{8\pi G}{3} \rho - \frac{1}{6} \left[\mathcal{P}^{(1)} + \mathcal{S}^{(1)} \right], \qquad (4.35a)$$

$$\frac{1}{a} \frac{d^2 a}{d\tau^2} = -\frac{4\pi G}{3} (\rho + 3p) + \frac{1}{3} \left[\mathcal{P}^{(1)} + \mathcal{P}^{(2)} + \mathcal{S}^{(2)} \right], \qquad (4.35b)$$

where the cosmic time τ is related to t via $d\tau = f(t)dt$, ρ and p are the homogeneous energy density and pressure defined using the averaged energy-momentum tensor, and the correlation terms are defined using the relations (with $H \equiv da/dt$),

$$\mathcal{P}^{(1)} = \frac{1}{f^2} \left[\langle \widetilde{\Gamma}^A_{0A} \widetilde{\Gamma}^B_{0B} \rangle - \langle \widetilde{\Gamma}^A_{0B} \widetilde{\Gamma}^B_{0A} \rangle - 6H^2 \right] , \qquad (4.36a)$$

$$\mathcal{S}^{(1)} = \langle \widetilde{g}^{JK} \rangle \left[\langle \widetilde{\Gamma}^A_{JB} \widetilde{\Gamma}^B_{KA} \rangle - \langle \widetilde{\Gamma}^A_{JA} \widetilde{\Gamma}^B_{KB} \rangle \right] , \qquad (4.36b)$$

$$\mathcal{P}^{(2)} + \mathcal{P}^{(1)} = -\frac{1}{f^2} \langle \widetilde{\Gamma}^A_{0A} \widetilde{\Gamma}^0_{00} \rangle - \langle \widetilde{g}^{JK} \rangle \langle \widetilde{\Gamma}^0_{JA} \widetilde{\Gamma}^A_{0K} \rangle + \frac{3H}{f^2} \left(\partial_t (\ln f) + H \right) , \qquad (4.36c)$$

$$\mathcal{S}^{(2)} = \frac{1}{f^2} \langle \widetilde{\Gamma}^A_{00} \widetilde{\Gamma}^0_{A0} \rangle + \langle \widetilde{g}^{JK} \rangle \langle \widetilde{\Gamma}^0_{J0} \widetilde{\Gamma}^A_{KA} \rangle . \qquad (4.36d)$$

Here $\widetilde{\Gamma}^i_{jk}$ is the bilocal extension of the Christoffel connection, and equations (4.35) are valid for an arbitrary choice of gauge on $\mathcal{M}$. We emphasize that averaging here refers to spatial averaging. Also $\langle \widetilde{g}^{JK} \rangle = G^{JK} = (1/\bar{a}^2)\delta^{JK}$, and the index 0 refers to the nonsynchronous time t. It can be shown that $\mathcal{P}^{(1)}$ and $\mathcal{P}^{(1)} + \mathcal{P}^{(2)}$ correspond to correlations of (the bilocal extensions of) the extrinsic curvature with itself and with the time derivative of the lapse function. $\mathcal{S}^{(1)}$ corresponds to correlations between the Christoffel symbols of the 3-geometry, and $\mathcal{S}^{(2)}$ to correlations of the spatial derivative of the lapse with itself and with the Christoffel symbols of the 3-geometry. This becomes especially clear when one works in the volume preserving gauge with $N\sqrt{h} = 1$, in which case one has

$$\mathcal{P}^{(1)} = \bar{a}^6 \left[\frac{2}{3} \left(\langle \frac{1}{h}\Theta^2 \rangle - \frac{1}{\bar{a}^6}({}^F\Theta^2) \right) - 2\langle \frac{1}{h}\sigma^2 \rangle \right] \; ; \; \frac{1}{\bar{a}^6}({}^F\Theta^2) = (3H)^2 , \qquad (4.37a)$$

$$\mathcal{S}^{(1)} = \frac{1}{\bar{a}^2}\delta^{AB} \left[\langle \, {}^{(3)}\Gamma^J_{AC} \, {}^{(3)}\Gamma^C_{BJ} \rangle - \langle \partial_A(\ln \sqrt{h})\partial_B(\ln \sqrt{h}) \rangle \right] , \qquad (4.37b)$$

$$\mathcal{P}^{(2)} = \bar{a}^6 \langle \frac{1}{h}\Theta^A_B\Theta^B_A \rangle - \frac{1}{\bar{a}^2}\delta^{AB}\langle \Theta_{AJ}\Theta^J_B \rangle, \qquad (4.37c)$$

$$\mathcal{S}^{(2)} = \bar{a}^6 \langle \frac{1}{h}h^{AB}\partial_A(\ln \sqrt{h})\partial_B(\ln \sqrt{h}) \rangle - \frac{1}{\bar{a}^2}\delta^{AB}\langle \partial_A(\ln \sqrt{h})\partial_B(\ln \sqrt{h}) \rangle . \qquad (4.37d)$$

This structure is closely related to that of the backreaction in Buchert's approach. There are also several differences, some of them somewhat subtle, between Buchert's and Zalaletdinov's approaches. We will not give a detailed comparison between the two formalisms here, referring the reader instead to Ref. [56] for a comprehensive discussion. Instead we will turn to an application of the spatial averaging limit of Zalaletdinov's approach, in the setting of linear cosmological perturbation theory.

4.5 Backreaction in cosmological perturbation theory

The perturbative context introduces some additional, purely technical subtleties into the problem. In order not to be bogged down by technical complications, let us start with some very simple order of magnitude estimates of the backreaction in a setting where the metric of the universe is of the perturbed FLRW form

$$ds^2 = a(\eta)^2 \left(-(1 + 2\varphi)d\eta^2 + (1 - 2\varphi)\gamma_{AB}dx^A dx^B \right) , \qquad (4.38)$$

where γ_{AB} is the metric of flat 3-space and the conformal time η is related to cosmic time τ via $d\tau = a(\eta)d\eta$. The structure of the backreaction is

$$\mathcal{C} \sim \langle \Gamma^2 \rangle - \langle \Gamma \rangle^2 , \qquad (4.39)$$

which, in the perturbative context with $\Gamma = \Gamma_{FLRW} + \delta\Gamma$ and $\langle \Gamma \rangle = \Gamma_{FLRW}$, reduces to

$$\mathcal{C} \sim \langle \delta\Gamma^2 \rangle , \qquad (4.40)$$

which leads to

$$\mathcal{C} \sim a^{-2}\langle \nabla\varphi \cdot \nabla\varphi \rangle , \qquad (4.41)$$

for the metric (4.38) if we ignore time derivatives of φ. Assuming a two component flat background consisting of cold dark matter (CDM) and radiation (known as standard CDM or sCDM), and taking the averaging to be an ensemble average over the initial conditions, it is not difficult to show that in the matter dominated era, this estimate leads to

$$\mathcal{C} \sim [10^{-4}H_0^2/a(\tau)^2] . \qquad (4.42)$$

The factor 10^{-4} arises from a product of the normalization of the initial power spectrum $A \sim 10^{-9}$, and the factor $(k_{eq}/H_0)^2 \sim 10^5$ which arises in the transfer function integral, where $k_{eq} = a_{eq}H(a_{eq})$ is the wavenumber corresponding to the radiation-matter equality scale. This indicates that at least for epochs around the last scattering epoch, the backreaction due to averaging was negligible.

The real situation is somewhat more complex than this simple calculation indicates. On the one hand, the time evolution of $a(\tau)$ is needed in order to solve the equations satisfied by the perturbations, as we effectively did above by assuming a form for $a(\tau)$. On the other hand, the evolution of the *perturbations* is needed to compute the correction terms $\mathcal{C}$. Until these corrections are known, the evolution of the scale factor cannot be determined; and until we know this evolution, we cannot solve for the perturbations.

To break this circle, we adopt an iterative procedure. We first compute a "zeroth iteration" estimate for the backreaction, by assuming a fixed standard background $a^{(0)}$ such as sCDM, evolve the perturbations and compute the time dependence of the objects $\mathcal{C}$, denoted $\mathcal{C}^{(0)}$. Now, using these *known* functions of time, we form a new estimate for the background $a^{(1)}$ using the modified equations, and hence calculate the "first iteration" estimate $\mathcal{C}^{(1)}$. This process can then be repeated, and is expected to converge as long as perturbation theory in the metric remains a valid approximation.

Another issue which comes up is the question of which *gauge* to choose in order to impose the condition $\langle \Gamma \rangle = \Gamma_{FLRW}$. It turns out that for consistency, this must be done in a volume preserving gauge, defined such that the metric determinant is a function of time alone (see Ref. [56]). The final results for the backreaction can be expressed in terms of the Newtonian potential appearing in Eqn. (4.7), and are given by

$$\mathcal{P}^{(1)} = \frac{1}{a^2} \left[2\langle (\varphi')^2 \rangle - \langle (\nabla_A \nabla_B \beta') (\nabla^A \nabla^B \beta') \rangle \right], \qquad (4.43a)$$

$$\mathcal{S}^{(1)} = -\frac{1}{a^2} \left[6\langle \partial_A \varphi \partial^A \varphi \rangle - \langle (\nabla_A \nabla_B \nabla_C \beta)(\nabla^A \nabla^B \nabla^C \beta) \rangle \right], \qquad (4.43b)$$

$$\mathcal{P}^{(1)} + \mathcal{P}^{(2)} = -\frac{2\mathcal{H}}{a^2} \langle (\nabla_A \nabla_B \beta)(\nabla^A \nabla^B \beta') \rangle, \qquad (4.43c)$$

$$\mathcal{S}^{(2)} = -\frac{1}{a^2} \left[\langle \partial^A \beta'' (\partial_A \varphi - \mathcal{H} \partial_A \beta') \rangle \right], \qquad (4.43d)$$

where $\mathcal{H} \equiv a'/a$ with a prime denoting a derivative with respect to conformal time, ∇_A is the covariant derivative compatible with γ_{AB}, and β is defined as the solution of $\nabla^2 \beta = -2\varphi$ with the condition that if $\varphi = 0$ then $\beta = 0$.

Typically in cosmology, initial conditions are set in the early universe during an inflationary phase, and are specified in terms of the statistical properties of a fluctuating quantum field. One then speaks of an ensemble average over many realizations of these fluctuations, and specifies e.g. their power spectrum in Fourier space. It is not hard to show, that performing such an ensemble average over and above our spatial average removes all reference to the scale of spatial averaging, and is equivalent to replacing the spatial average by the ensemble average to begin with, and we will do this in what follows. This is valid however only in the situation where there are no fluctuations at arbitrarily large length scales, since in the presence of such fluctuations the averaging condition $\langle \Gamma \rangle = \Gamma_{\text{FLRW}}$ loses meaning (in such a situation it would be impossible to isolate the background from the perturbation by an averaging operation on any finite length scale, see below).

It is convenient to define the transfer function (in Fourier space) $\Phi_k(\eta)$ via the relation

$$\varphi_{\vec{k}}(\eta) = \varphi_{\vec{k}i}\Phi_k(\eta) \,, \tag{4.44}$$

and define the power spectrum of the initial fluctuations $\varphi_{\vec{k}i}$ via

$$\langle \varphi_{\vec{k}_1 i}\varphi^*_{\vec{k}_2 i} \rangle = (2\pi)^3 \delta^{(3)}(\vec{k}_1 - \vec{k}_2) P_{\varphi i}(k_1) \,. \tag{4.45}$$

The correlation scalars (4.43) can then be written as

$$\mathcal{P}^{(1)} = -\frac{2}{a^2}\int \frac{dk}{2\pi^2}k^2 P_{\varphi i}(k)\,(\Phi'_k)^2 \,, \tag{4.46a}$$

$$\mathcal{S}^{(1)} = -\frac{2}{a^2}\int \frac{dk}{2\pi^2}k^2 P_{\varphi i}(k)\,(k^2\Phi_k^2) \,, \tag{4.46b}$$

$$\mathcal{P}^{(1)} + \mathcal{P}^{(2)} = -\frac{8\mathcal{H}}{a^2}\int \frac{dk}{2\pi^2}k^2 P_{\varphi i}(k)\,(\Phi_k\Phi'_k) \,, \tag{4.46c}$$

$$\mathcal{S}^{(2)} = -\frac{2}{a^2}\int \frac{dk}{2\pi^2}k^2 P_{\varphi i}(k)\Phi''_k\left(\Phi_k - \frac{2\mathcal{H}}{k^2}\Phi'_k\right) \,. \tag{4.46d}$$

These expressions highlight the problem of having a finite amplitude for fluctuations at arbitrarily large length scales ($k \to 0$). As a concrete example, consider the frequently discussed Harrison-Zel'dovich scale invariant spectrum [58] which satisfies the condition

$$k^3 P_{\varphi i}(k) = \text{constant}\,. \tag{4.47}$$

Eqns. (4.46) now show that if the transfer function $\Phi_k(\eta)$ has a finite time derivative at large scales (as it does in the standard scenarios), then the correlation objects $\mathcal{P}^{(1)}$, $\mathcal{P}^{(2)}$ and $\mathcal{S}^{(2)}$ all diverge due to contributions from the $k \to 0$ regime. This demonstrates the importance of having an initial power spectrum in which the amplitude dies down sufficiently rapidly on large length scales (which is a known issue, see Ref. [59]). Keeping this in mind, we shall concentrate on initial power spectra which display a long wavelength cutoff. Models of inflation leading to such power spectra have been discussed in the literature [60], and more encouragingly, analyses of WMAP data seem to indicate that such a cutoff in the initial power spectrum is in fact realized in the universe [61].

The results of a numerical calculation using the two component sCDM model mentioned earlier, are shown in figure 4.1, where all functions are normalized by the Hubble parameter $H^2(a)$. We see that this zeroth iteration estimate in fact gives a negligible contribution.

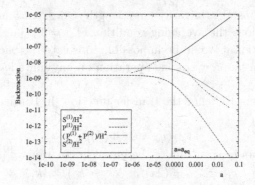

Fig. 4.1 The backreaction for the sCDM model, normalized by $H^2(a)$. $\mathcal{S}^{(1)}$, $\mathcal{P}^{(1)}$ and $\mathcal{S}^{(2)}$ are negative definite and their magnitudes have been plotted. The vertical line marks the epoch of matter radiation equality $a = a_{eq}$.

4.5.1 *Lessons from linear theory*

We see that the dominant contribution to the backreaction at late times, is due to a curvature-like term $\sim a^{-2}$, as expected from our simple estimate above. In order to obtain a correction which grows faster than this, we need a nonstandard evolution of the metric potential φ, which can only happen if the *scale factor* evolves very differently from the sCDM model, which in turn would require a significant contribution from the backreaction. The same circle of dependencies as before, now implies that *as long as the metric is perturbed FLRW*, the backreaction appears to be dynamically suppressed. Secondly, as figures 4.2 and 4.3 show, scales which are approaching nonlinearity, *do not* contribute significantly to the backreaction, which is a consequence of the suppression of small scale power by the transfer function. We will return to this point below when discussing the backreaction during epochs of nonlinear structure formation.

Fig. 4.1 shows that in the absence of a cosmological constant, the backreaction after a single iteration, tracks the radiation density in the radiation dominated era, and essentially behaves like a curvature term in the matter dominated era. Two issues arise from this behavior. Denote the corrections to the Friedmann equation and the acceleration equation (respectively the first and second equations in (4.17)) as $\mathcal{C}_F$ and $\mathcal{C}_{acc}$ respectively. Then firstly, we find that in the radiation dominated era, although both $\mathcal{C}_F$ and $\mathcal{C}_{acc}$ behave like $\sim a^{-4}$, their numerical coefficients do not combine so as to yield an effective fluid with a conserved energy-momentum tensor. Since the backreaction must now necessarily couple to the background radiation

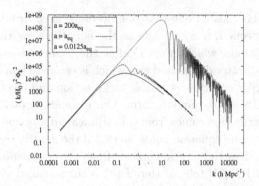

Fig. 4.2   The dimensionless integrand of $\mathcal{S}^{(1)}$, namely the function $(k/H_0)^2 \Phi_k^2$, at three sample values of the scale factor. The function dies down rapidly for large k, with the value at some k being progressively smaller with increasing scale factor. The declining behavior of the curves for $a = a_{eq}$ and $a = 200 a_{eq}$ extrapolates to large k.

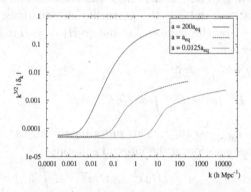

Fig. 4.3   The dimensionless CDM density contrast. Together with figure 4.3 this shows that nonlinear scales do not impact the backreaction integrals significantly.

density, this points to a very tiny gravitationally induced correction in the equation of state for radiation. This effect can be traced back essentially to the presence of a small but non-zero correlation 3-form and 4-form, arising from higher order perturbative effects. In the matter dominated era, at least in the "zeroth" iteration, this effect seems to be highly suppressed since we now have $\mathcal{C}_F \sim a^{-2}$ and $|\mathcal{C}_{acc}| \ll |\mathcal{C}_F|$, which is approximately consistent with conservation of the effective energy-momentum tensor.

The second issue concerns what happens at higher iterations, and is important from the point of view of obtaining a convergent answer for the backreaction. The basic cycle that one needs to keep in mind is that the backreaction affects H^2, which affects the equations for the density and metric perturbations, which in turn define the backreaction. Consider the situation in the matter dominated era, which is easier to handle since firstly only one term $\mathcal{S}^{(1)}$ contributes to the backreaction and secondly the linear PT solution has a simple analytic form. Our estimate in (4.42) showed that most of the contribution comes from (quasi)linear subhorizon scales $k \sim k_{eq}$ for which the Poisson equation holds, so that if the density contrast behaves like $\delta_k \sim D(a)$ then the metric transfer function behaves like $\Phi_k \sim D/a$. Standard linear PT [11] tells us that $D(a)$ is the so-called growth function which can be written as $D \sim E \int da/(aE)^3$ upto some numerical coefficient, where $E \equiv H/H_0$. An analysis similar to the one leading to Eqn. (4.42) then shows that we should expect $\mathcal{C}_F \sim a^{-2}(D/a)^2$ at late times. For a flat universe without a cosmological constant, $D(a) = a$ and we recover the single iteration result that we have been discussing so far.

The crucial thing to note is that since the backreaction affects only the background equations and *not* the perturbation equations, $D(a)$ is completely determined by the Hubble parameter $H(a) = H_0 E(a)$, so that at any iteration i we will have

$$(E^{(i+1)})^2 = \Omega_m^{(i+1)} a^{-3} + \epsilon_{\text{bkrxn}}^{(i)} a^{-2}(D^{(i)}/a)^2, \tag{4.48a}$$

$$D^{(i)}(a) \sim E^{(i)} \int da/(aE^{(i)})^3, \tag{4.48b}$$

where we expect $\epsilon_{\text{bkrxn}}^{(i)} \sim 10^{-4}$. This immediately suggests that the limit of this series is the solution of the integral equation

$$E^2 = \Omega_m a^{-3} + \epsilon_{\text{bkrxn}} a^{-2}(D/a)^2 \;\; ; \;\; D(a) \sim E \int da/(aE)^3. \tag{4.49}$$

This equation can in principle be solved perturbatively by exploiting the smallness of the parameter ϵ_{bkrxn}, and we expect the solution to be close to the "zeroth" iteration answer $D \sim a$. To understand why, notice that at the zeroth iteration we found $\mathcal{S}^{(1)}$ to be negative, so that the first iteration Hubble parameter $E^{(1)}(a)$ is effectively that of an *open* universe with a small negative curvature. Standard analysis shows that the growth factor in an open universe is suppressed compared to that in a flat matter dominated one, and hence the Hubble parameter at the *second* iteration $E^{(2)}(a)$ will have a slightly smaller contribution from the backreaction than $E^{(1)}(a)$. This will correspondingly slightly *enhance* the contribution of the backreaction to $E^{(3)}(a)$ over the contribution to $E^{(2)}(a)$, and so on until the solution converges.

This convergent solution will, like the radiation dominated case, mildly violate the conservation criterion Eqn. (4.42). Further, the analysis above generalizes to the case when the cosmological constant is nonzero. In this case the late time growth factor is suppressed compared to the EdS case even at the zeroth iteration [11], and the convergent solution will violate the conservation criterion by an amount comparable to the backreaction itself. What is important however is that in *all* cases, the backreaction as well as the violation of matter conservation remain negligibly small, approximately at the level of one part in 10^4.

4.5.2 *The nonlinear regime*

The previous analysis ignored all contributions from scales which have become fully nonlinear in the matter density contrast at late times. The reasoning was that these scales are not expected to contribute significantly to the backreaction due to a suppression in the transfer function Φ_k. Let us now ask whether one can make meaningful statements concerning the backreaction during epochs of nonlinear structure formation, when matter density contrasts become very large and perturbation theory in the matter variables has broken down. We begin by considering some order of magnitude estimates.

4.5.2.1 *Dimensional arguments, and why they fail*

Let us start with the assumption that although the matter perturbations are large, one can still expand the *metric* as a perturbation around FLRW. We are looking for either self-consistent solutions using this assumption, or any indication that this assumption is not valid. Given that the metric has the form (4.7), the relevant gravitational equation at late times and at length scales small comparable to H^{-1}, is the Poisson equation given by

$$\frac{1}{a^2}\nabla^2\varphi = 4\pi G\bar{\rho}\delta \,, \qquad (4.50)$$

where $\delta \equiv (\rho(t, \vec{x})/\bar{\rho}(t) - 1)$ is the density contrast of CDM[3]. As before, we can estimate the dominant backreaction component to be $\mathcal{C} \sim a^{-2}\langle \nabla\varphi \cdot \nabla\varphi \rangle$.

Now, for an over/under-density of physical size R, treating $a^{-1}\nabla \sim R^{-1}$

---

[3]We are only worried about the small, sub-Hubble scales, since larger scales are well described by linear theory where we know the form of the backreaction.

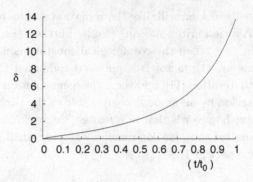

Fig. 4.4 Density contrast in the LTB toy model, at $r = 8.35\,\text{Mpc}$. $t_0 = 2/3H_0 \approx 9\,\text{Gyr}$.

and $G\bar{\rho} \sim H^2$ on dimensional grounds, we have

$$|\varphi| \sim (HR)^2\, |\delta| \,. \tag{4.51}$$

For voids, we can set $\delta \sim -1$, and then $\mathcal{C} \sim H^2(HR)^2 \ll H^2$, since we have assumed $HR \ll 1$. This shows that sub-Hubble underdense voids are expected to give a negligible backreaction. For overdense regions we need to be more careful, since here δ can grow very large. In a typical spherical collapse scenario, the following relations hold,

$$R \sim (1 - \cos u)r \; ; \; H^{-1} \sim (G\bar{\rho})^{-1/2} \sim t \sim H_0^{-1}(u - \sin u)\,, \tag{4.52}$$

$$G\rho \sim \frac{(H_0 r)^2}{R^2 R'} \sim \frac{H_0^2}{(1 - \cos u)^3} \; ; \; \delta \sim (\rho/\bar{\rho}) \sim \frac{(u - \sin u)^2}{(1 - \cos u)^3}\,, \tag{4.53}$$

which lead to

$$|\varphi| \sim \frac{(H_0 r)^2}{(1 - \cos u)} \; ; \; \mathcal{C} \sim H^2 \left[(H_0 r)^2 \frac{(u - \sin u)^2}{(1 - \cos u)^4} \right]\,. \tag{4.54}$$

It would therefore appear that at late enough times, the perturbative expansion in the metric breaks down with $|\varphi| \sim 1$, and the backreaction grows large $|\mathcal{C}| \sim 1$. However, the crucial question one needs to answer is the following : Is this situation actually realized in the universe, or are we simply taking these models too far? We claim that perturbation theory in the metric *does not* break down at late times, since *observed peculiar velocities remain small*. The spherical collapse model is not a good approximation when *model* peculiar velocities in the collapsing phase grow large. To support this claim, we will work with an exact toy model of spherical collapse.

4.5.3   Calculations in an exact model

The model we consider was used in Ref. [62], and can be summarized as follows. The matter content of the model is spherically symmetric pressureless "dust", and hence the relevant exact solution is the Lemaître-Tolman-Bondi (LTB) metric given by

$$ds^2 = -dt^2 + \frac{R'^2 dr^2}{1 - k(r)r^2} + R^2 d\Omega^2 . \tag{4.55}$$

Here t is the proper time measured by observers with fixed coordinate r, which is comoving with the dust. $R(t, r)$ is the physical area radius of the dust shell labelled by r, and satisfies the equation $\dot{R}^2 = 2GM(r)/R - k(r)r^2$. Here $M(r)$ is the mass contained inside each comoving shell, and a dot denotes a derivative with respect to the proper time t. The energy density of dust measured by an observer comoving with it satisfies the equation $\rho(t, r) = M'(r)/4\pi R^2 R'$, where the prime now denotes a derivative with respect to the LTB radius r.

Initial conditions are set at a scale factor value of $a_i = 10^{-3}$, and are chosen such that the initial situation describes an FLRW expansion with a perturbative central overdensity out to radius $r = r_*$, surrounded by a perturbative underdensity out to radius $r = r_v$, with appropriately chosen values for the various parameters in the model (see Table 1 of Ref. [62]). Figure 4.4 shows the evolution of the overdensity contrast in the central region. Clearly, at late times the situation is completely nonlinear. Nevertheless, it can be shown that a coordinate transformation in this model can bring its metric to the form (4.7), *provided* one has $|av| \ll 1$ where $v = \partial_t(R/a)$. Physically v is the "comoving" peculiar velocity. The metric potential has the expression $\varphi = -\dot{\xi}^0 + (1/2)(av)^2$, where ξ^0 is obtained by integrating $\xi^{0\prime} = avR'$. A numerical calculation shows that av and hence the metric potentials do in fact remain small for the entire evolution, for *this* model. Further, the infall peculiar velocity can only become large if the true infall velocity $\dot{R}$ is large, in which case the specific background chosen to define the peculiar velocity, becomes irrelevant (since $HR \ll 1$). Hence, the fact that relativistic infall velocities are *not* observed in real clusters etc., leads us to expect very generally that the perturbed FLRW form for the metric should in fact be recoverable even at late times.

Finally, figure 4.5 shows the dominant contribution to the backreaction in the toy model [63]. There is a significant departure from a curvature-like behavior, due to evolution of the metric potentials. More importantly, the maximum value of the backreaction here is $\sim 10^{-6} H^2$, as opposed to

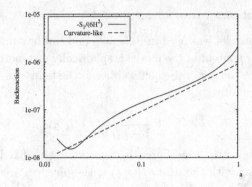

Fig. 4.5 The evolution of $|\mathcal{S}^{(1)}|/6H^2$. Also shown is a hypothetical curvature-like correction, evolving like $\sim a^{-2}$.

$\sim 10^{-4}H^2$ as seen in the linear theory. This can be understood by noting that the inhomogeneity of our toy model is only on relatively small, nonlinear scales ($\lesssim 20h^{-1}\mathrm{Mpc}$), and the value of the backreaction is therefore consistent with our earlier observation that nonlinear scales contribute negligibly to the total backreaction. To conclude this section, we have seen that as long as the metric has the perturbed FLRW form, the backreaction remains small. Further, there are strong reasons to expect that the metric remains a perturbation around FLRW even at late times during nonlinear structure formation, a claim that is supported by our toy model calculation. It should be possible to test this claim in N-body simulations as well. It appears therefore, that backreaction cannot explain the observed acceleration of the universe.

4.6 Conclusions

Backreaction as an explanation for the late time cosmic acceleration would have truly been the most conservative solution to the dark energy problem. Not only would it have resolved the discrepancy between observed data and what is generally considered to be "ordinary physics", but more importantly it would have obviated the need for statements such as "We do not understand what 70% of the universe is made of"[4]. Needless to say,

---

[4]Including dark matter would take this number up to approximately 95%. If we further take into account the fact that the only component which we directly measure with great precision is the CMBR, then one might say that we truly understand only $\sim 10^{-4}$ of the universe!

this approach has captured the imagination of many cosmologists, and the (possibly incomplete) list of references cited in the Introduction is testimony to this fact. Due to the technically challenging nature of the problem however, it is very important to proceed systematically and rigorously while determining the size and nature of the effects of backreaction. This is especially true since order of magnitude estimates on the one hand indicate that the effect can never be large [28], while simple toy models indicate exactly the opposite [38]. As we have seen however, when the calculation is performed in a reliable and self-consistent manner, the results agree with the conventional wisdom. Backreaction from averaging cannot therefore solve the dark energy problem.

The idea of using inhomogeneities to explain the dark energy problem has generated a flurry of research in the backreaction problem in recent years, as we saw earlier (see also Ref. [64]). It is important to also mention another approach which has gained popularity in this context, namely that of ascribing the dark energy phenomenon to light propagation effects in an inhomogeneous universe [65]. We conclude with a brief discussion of this approach.

4.6.1 *The "Special Observer" assumption*

The central idea here is that light propagation through an inhomogeneous underdensity or "void" can be significantly different from that in a homogeneous space. In fact, it is possible to show that luminosity distance data from supernovae can always be fit by modelling ourselves as observers in a void with a suitable density profile. Typically however, the (usually spherical) voids invoked for this purpose are very large (in the range of $\sim 200h^{-1}\text{Mpc}$ to $\sim 1h^{-1}\text{Gpc}$ in diameter), and are difficult to reconcile with the typical sizes of voids seen in galaxy surveys, which are in the range of $30\text{-}50h^{-1}\text{Mpc}$, with some "supervoids" reaching $\sim 100h^{-1}\text{Mpc}$ [66]. Nevertheless, this idea has been rather vigorously investigated in the last several years [67]. Unlike the backreaction issue which requires mainly theoretical work, a detailed description of the inhomogeneous universe belongs squarely in the regime of observational cosmology [68]. Due to the obvious observational difficulties involved in such a program (for example due to the lack of homogeneous samples of galaxy data), this approach at present is largely restricted to being an exercise in building toy models of the local large scale structure (although see the last three references in [67] for recent constraints). As a final comment on this topic, we note that this "non-Copernican" approach (even at the level of building toy models) is amenable to observational verification or disproof in the coming generation of surveys, as pointed out by Ref. [69].

To conclude, backreaction from inhomogeneities cannot solve the dark energy problem. Void-like inhomogeneities, while having the potential to solve this problem, await further observational evidence. And for now, we still do not understand what (at least) 70% of the universe is made of. The future continues to hold significant challenges for cosmology and theoretical physics.

References

[1] For a history of the Big Bang model, see J -P Luminet, arXiv:0704.3579, (2007).

[2] S Weinberg, *Gravitation and Cosmology* (John Wiley and Sons, Inc., 1972).

[3] G F Smoot *et al.*, *Astrophys. J.* **396**, L1 (1992).

[4] D N Spergel *et al.*, *Astrophys. J. Suppl.* **170**, 377 (2007) [arXiv:astro-ph/0603449].

[5] J A Peacock *et al.*, *Nature* **410**, 169 (2001) [arXiv:astro-ph/0103143].

[6] C Hikage *et al.* (SDSS collaboration), *Publ. Astron. Soc. Jap.* **55**, 911 (2003) [arXiv:astro-ph/0304455];
J Yadav *et al.*, *Mon. Not. Roy. Astron. Soc.* **364**, 601 (2005) [arXiv:astro-ph/0504315].

[7] See, e.g.,
D W Hogg *et al.*, *Astrophys. J.* **642**, 54 (2005), [arXiv:astro-ph/0411197];
L Pietronero and F S Labini, *AIP Conf. Proc.* **822**, 294 (2006), [arXiv:astro-ph/0406202];
M Joyce *et al.*, *Astron. & Astrophys.* **443**, 11 (2005), [arXiv:astro-ph/0501583];
M Kerscher, J Schmalzing, T Buchert and H Wagner, *Astron. & Astrophys.* **333**, 1 (1998), [arXiv:astro-ph/9704028];
M. Kerscher *et al.*, *Astron. & Astrophys.* **373**, 1 (2001), [arXiv:astro-ph/0101238];

[8] F Sylos Labini, N L Vasilyev and Y V Baryshev, *Europhys. Lett.* **85**, 29002 (2009) [arXiv:0812.3260];
F Sylos Labini et al., arXiv:0805.1132 (2008).

[9] V J Martinez, arXiv:0804.1536, to appear in *"Data Analysis in Cosmology"*, *Lecture Notes in Physics* (2008), eds. V J Martinez, et al., Springer-Verlag.

[10] P J E Peebles, *Principles of Physical Cosmology*, Princeton Univ. Press, New Jersey (1993).

[11] S Dodelson, *Modern Cosmology*, Academic Press, San Diego (2003).

[12] V F Mukhanov, H A Feldman and R H Brandenberger, *Phys. Rept.* **215**, 203 (1992).

[13] W H Press and P Schechter, *Astrophys. J.* **187**, 425 (1974); R K Sheth and G Tormen, *Mon. Not. Roy. Astron. Soc.* **329**, 61 (2002) [arXiv:astro-ph/0105113].

[14] F Bernardeau, S Colombi, E Gaztanaga and R Scoccimarro, *Phys. Rept.* **367**, 1 (2002) [arXiv:astro-ph/0112551].

[15] M Trenti and P Hut, arXiv:0806.3950 (2008), invited refereed review for the Scholarpedia Encyclopedia of Astrophysics, available online at http://www.scholarpedia.org/article/N-body_simulations_(gravitational)

[16] G F R Ellis, in *General Relativity and Gravitation* (D. Reidel Publishing Co., Dordrecht, 1984), Eds. B. Bertotti *et al.*

[17] A Riess et al., *Astrophys. J.* **607**, 664 (2004) [arXiv:astro-ph/0402512]; M Seikel and D J Schwarz, *JCAP* **02**(2008)007 [arXiv:0711.3180]; E Mörtsell and C Clarkson, arXiv:0811.0981 (2008), JCAP in press.

[18] For a comprehensive review, see E J Copeland, M Sami and S Tsujikawa, *Int. J. Mod. Phys.* **D15**, 1753 (2006), [arXiv:hep-th/0603057].

[19] A Krasiński, *Inhomogeneous Cosmological Models*, (Cambridge Univ. Press, 1997).

[20] R A Isaacson, *Phys. Rev.* **166**, 1272 (1968).

[21] D R Brill and J B Hartle, *Phys. Rev.* **135**, B271 (1964).

[22] T W Noonan, *Gen. Rel. Grav.* **16**, 1103 (1984).

[23] T Futamase, *Phys. Rev. Lett.* **61**, 2175 (1988).

[24] T Futamase, *Phys. Rev.* **D53**, 681 (1996).

[25] J P Boersma, *Phys. Rev.* **D57**, 798 (1998), [arXiv:gr-qc/9711057].

[26] M Kasai, *Phys. Rev. Lett.* **69**, 2330 (1992).

[27] W R Stoeger, G F R Ellis and C Hellaby, *Mon. Not. Roy. Astron. Soc.* **226**, 373 (1987); M Carfora and K Piotrkowska, *Phys. Rev.* **D52**, 4393 (1995) [arXiv:gr-qc/9502021]; N Mustapha, B A Bassett, C Hellaby and G F R Ellis, *Class. Quant. Grav.* **15**, 2363 (1998) [arXiv:gr-qc/9708043];

A Krasiński, *Inhomogeneous Cosmological Models*, Cambridge Univ. Press (1997);
M Reiris, *Class. Quant. Grav.* **25**, 085001 (2008) [arXiv:0709.0770];
C Anastopoulos, arXiv:0902.0159 (2009).

[28] A Ishibashi and R M Wald, *Class. Quant. Grav.* **23**, 235 (2006) [arXiv:gr-qc/0509108].

[29] P Martineau and R H Brandenberger, arXiv:gr-qc/0509108 (2005).

[30] L R Abramo, R H Brandenberger and V F Mukhanov, *Phys. Rev.* **D56**, 3248 (1997) [arXiv:gr-qc/9704037].

[31] E Barausse, S Matarrese and A Riotto, *Phys.Rev.* **D71**, 063537 (2005) [arXiv:astro-ph/0501152].

[32] E W Kolb, S Matarrese, A Notari and A Riotto, arXiv:hep-th/0503117 (2005).

[33] C M Hirata and U Seljak, *Phys. Rev.* **D72**, 083501 (2005) [arXiv:astro-ph/0503582];
G Geshnizjani, D J H Chung and N Afshordi, *Phys. Rev.* **D72** 023517 (2005) [arXiv:astro-ph/0503553];
E E Flanagan, *Phys. Rev.* **D71**, 103521 (2005) [arXiv:hep-th/0503202];
S Räsänen, *Class. Quant. Grav.* **23**, 1823 (2006) [arXiv:astro-ph/0504005];
M F Parry, *JCAP* **0606**:016 (2006) [arXiv:astro-ph/0605159];
N Kumar and E E Flanagan, *Phys. Rev.* **D78**, 063537 (2008) [arXiv:0808.1043].

[34] W R Stoeger, A Helmi, D F Torres, *Int. J. Mod. Phys.* **D16**, 1001 (2007) [arXiv:gr-qc/9904020];
E A Calzetta, B L Hu and F D Mazzitelli, *Phys. Rept.* **352**, 459 (2001) [arXiv:hep-th/0102199];
C Wetterich, *Phys. Rev.* **D67**, 043513 (2003) [arXiv:astro-ph/0111166];
E R Siegel and J N Fry, *Astrophys. J.* **628**, L1 (2005) [arXiv:astro-ph/0504421];
A Gruzinov, M Kleban, M Porrati and M Redi, *JCAP* **0612**:001 (2006) [arXiv:astro-ph/0609553];
M Gasperini, G Marozzi and G Veneziano, arXiv:0901.1303 (2009).

[35] A Notari, *Mod. Phys. Lett.* **A21**, 2997 (2006) [arXiv:astro-ph/0503715]

[36] K Van Acoleyen, *JCAP* **0810**:028 (2008) [arXiv:0808.3554].

[37] N Li and D J Schwarz, *Phys. Rev.* **D76**, 083011 (2007) [arXiv:gr-qc/0702043].

[38] S Räsänen, *JCAP* **0611**:003 (2006) [arXiv:astro-ph/0607626].

[39] D L Wiltshire, *New J. Phys.* **9**, 377 (2007) [arXiv:gr-qc/0702082]; *Int.*

J. Mod. Phys. **D17**, 641 (2008) [arXiv:0712.3982]; *Phys. Rev. Lett.* **99**, 251101 (2007) [arXiv:0709.0732];

B M Leith, S C Cindy Ng, D L Wiltshire, *Astrophys. J.* **672**, L91 (2008) [arXiv:0709.2535].

[40] J Kwan, M J Francis and G F Lewis, arXiv:0902.4249 (2009).

[41] T Buchert, *Gen. Rel. Grav.* **32**, 105 (2000) [arXiv:gr-qc/9906015].

[42] T Buchert, *Gen. Rel. Grav.* **33**, 1381 (2000) [arXiv:gr-qc/0102049].

[43] T Buchert and M Carfora, *Class. Quant. Grav.* **19**, 6109 (2002) [arXiv:gr-qc/0210037]; *Phys. Rev. Lett.* **90**, 031101 (2003) [arXiv:gr-qc/0210045].

[44] R M Zalaletdinov, *Gen. Rel. Grav.* **24**, 1015 (1992); *Gen. Rel. Grav.* **25**, 673 (1993).

[45] M Mars and R M Zalaletdinov, *J. Math. Phys.*, **38**, 4741, (1997) [dg-ga/9703002].

[46] R M Zalaletdinov, *Bull. Astron. Soc. India*, **25**, 401, (1997) [gr-qc/9703016].

[47] T Buchert and J Ehlers, *Astron. & Astrophys.* **320**, 1 (1997) [arXiv:astro-ph/9510056].

[48] T Buchert, M Kerscher and C Sicka, *Phys. Rev.* **D62**, 043525 (2000) [arXiv:astro-ph/9912347].

[49] D Palle, *Nuovo Cim.* **117B**, 687 (2002) [arXiv:astro-ph/0205462];

S Räsänen, *JCAP* **0402**:003 (2004) [arXiv:astro-ph/0311257]; *JCAP* **0411**:010 (2004) [arXiv:gr-qc/0408097]; *JCAP* **0804**:026 (2008) [arXiv:0801.2692]; *JCAP* **0902**:011 (2009) [arXiv:0812.2872];

G F R Ellis and T Buchert, *Phys. Lett.* **A347**, 38 (2005) [arXiv:gr-qc/0506106];

E W Kolb, S Matarrese and A Riotto, *New J. Phys.* **8**, 322 (2006) [arXiv:astro-ph/0506534];

Y Nambu and M Tanimoto, arXiv:gr-qc/0507057 (2005);

J D Barrow and C G Tsagas, *Class. Quant. Grav.* **24**, 1023 (2007) [arXiv:gr-qc/0609078];

A E Romano, *Phys. Rev.* **D75**, 043509 (2007) [arXiv:astro-ph/0612002];

M-N Celerier, *New Adv. Phys.* **1**, 29 (2007) [arXiv:astro-ph/0702416];

T Buchert, *Gen. Rel. Grav.* **40**, 467 (2008) [arXiv:0707.2153];

T Buchert and M Carfora, *Class. Quant. Grav.* **25**, 195001 (2008) [arXiv:0803.1401];

V F Cardone and G Esposito, arXiv:0805.1203 (2008);

R A Sussman, arXiv:0807.1145 (2008);

K Bolejko and L Andersson, *JCAP* **0810**:003 (2008) [arXiv:0807.3577];
K Bolejko, arXiv:0808.0376 (2008);
J Larena, arXiv:0902.3159 (2009).

[50] J Behrend, I A Brown and G Robbers, *JCAP* **0801**:013 (2008) [arXiv:0710.4964].

[51] T Buchert, J Larena and J-M Alimi, *Class. Quant. Grav.* **23**, 6379 (2006) [arXiv:gr-qc/0606020].

[52] V Marra, arXiv:0803.3152 (2008), PhD Thesis;
E W Kolb, V Marra and S Matarrese, *Phys. Rev.* **D78**, 103002 (2008) [arXiv:0807.0401].

[53] N Li and D J Schwarz, *Phys. Rev.* **D78**, 083531 (2008) [arXiv:0710.5073];
J Larena, J-M Alimi, T Buchert, M Kunz and P-S Corasaniti, arXiv:0808.1161 (2008);
E Rosenthal and E E Flanagan, arXiv:0809.2107 (2008);
K Bolejko, A Kurek and M Szydlowski, arXiv:0811.4487 (2008).

[54] A A Coley, N Pelavas and R M Zalaletdinov, *Phys. Rev. Lett.*, **95**, 151102, (2005) [gr-qc/0504115];
A A Coley and N Pelavas, *Phys. Rev.* **D74**, 087301, (2006) [astro-ph/0606535]; *Phys. Rev.* **D75**, 043506, (2007) [gr-qc/0607079];
R J Van Den Hoogen, *Gen. Rel. Grav.* **40**, 2213 (2008) [arXiv:0710.1823];
A A Coley, arXiv:0812.4565 (2008).

[55] L D Landau and E M Lifshitz, *The Classical Theory of Fields*, 4th rev. English ed., Butterworth-Heinemann, Oxford (1975).

[56] A Paranjape, PhD thesis, arXiv:0906.3165 (2009).

[57] C W Misner, K S Thorne and J A Wheeler, *Gravitation*, (W H Freeman and Co., New York, 1970).

[58] E R Harrison, *Phys. Rev.* **D1**, 2726 (1970);
Y B Zel'dovich, *Mon. Not. Roy. Astron. Soc.* **160**, 1P (1972).

[59] A R Liddle and D H Lyth, *Cosmological Inflation and Large Scale Structure*, Cambridge Univ. Press (2000).

[60] See, e.g.
A Vilenkin and L H Ford, *Phys. Rev.* **D26**, 1231 (1982);
J Silk and M S Turner, *Phys. Rev.* **D35**, 419 (1987);
L A Kofman, A D Linde, *Nucl. Phys.* **B282**, 555 (1987).

[61] A Shafieloo and T Souradeep, *Phys. Rev.* **D70**, 043523 (2004) [arXiv:astro-ph/0312174];
D Tocchini-Valentini, Y Hoffman and J Silk, *Mon. Not. Roy. Astron.*

Soc. **367**, 1095 (2006) [arXiv:astro-ph/0509478];
See, however,
L Verde and H V Peiris [arXiv:0802.1219 [astro-ph]] (2008).

[62] A Paranjape and T P Singh, *JCAP* **03**, 023 (2008) [arXiv:0801.1546].

[63] A Paranjape and T P Singh, *Phys. Rev. Lett.* **101**, 181101 (2008) [arXiv:0806.3497].

[64] J W Moffat, *JCAP* **0605**:001 (2006) [arXiv:astro-ph/0505326];
T Mattsson and M Ronkainen, *JCAP* **0802**:004 (2008) [arXiv:0708.3673];
G M Hossain, arXiv:0709.3490 (2007);
T Mattsson, arXiv:0711.4264 (2007).

[65] J Kristian and R K Sachs, *Astrophys. J* **143**, 379 (1966);
C C Dyer and R C Roeder, *Astrophys. J* **174**, L115 (1972); *Astrophys. J* **189**, 167 (1974).

[66] F Hoyle and M S Vogeley, *Astrophys. J.* **566**, 641 (2002) [arXiv:astro-ph/0109357]; *Astrophys. J.* **607**, 751 (2004) [arXiv:astro-ph/0312533];
M S Vogeley *et al.*, in *Proc. IAU Colloquium No. 195* (2004), A Diaferio, ed.
S G Patiri, *et al.*, *Mon. Not. R. Astr. Soc.* **369**, 335 (2006) [arXiv:astro-ph/0506668];
A V Tikhonov and I D Karachentsev, *Astrophys. J.* **653**, 969 (2006) [astro-ph/0609109].

[67] M Sasaki, *Mon. Not. Roy. Astron. Soc.* **228**, 653 (1987);
N Sugiura, K Nakao and T Harada, *Phys. Rev.* **D60**, 103508 (1999);
M-N Celerier, *Astron. & Astrophys.* **353**, 63 (2000) [arXiv:astro-ph/9907206];
K Tomita, *Mon. Not. Roy. Astron. Soc.* **326**, 287 (2001) [arXiv:astro-ph/0011484];
H Iguchi, T Nakamura and K Nakao, *Prog. of Theo. Phys.* **108**, 809 (2002);
W Godlowski, J Stelmach and M Szydlowski, *Class. Quant. Grav.* **21**, 3953 (2004) [arXiv:astro-ph/0403534];
H Alnes, M Amazguioui and O Gron, *Phys. Rev.* **D73**, 083519 (2006) [arXiv:astro-ph/0512006];
R A Vanderveld, E E Flanagan and I Wasserman, *Phys. Rev.* **D74**, 023506 (2006) [arXiv:astro-ph/0602476]; *Phys. Rev.* **D76**, 083504 (2007) [arXiv:0706.1931];
C H Chuang, J A Gu and W Y P Hwang, *Class. Quant. Grav.* **25**, 175001 (2008) [arXiv:astro-ph/0512651];

K Bolejko, *PMC Phys.* **A2**, 1 (2008) [arXiv:astro-ph/0512103];

P Apostolopoulos, N Brouzakis, N Tetradis and E Tzavara, *JCAP* **0606**:009 (2006) [arXiv:astro-ph/0603234];

D Garfinkle, *Class. Quant. Grav.* **23**, 4811 (2006) [arXiv:gr-qc/0605088];

T Biswas, R Mansouri and A Notari, *JCAP* **0712**:017 (2007) [arXiv:astro-ph/0606703];

H Alnes and M Amazguioui, *Phys. Rev.* **D74**, 103520 (2006) [arXiv:astro-ph/0607334];

K Enqvist and T Mattsson, *JCAP* **0702**:019 (2007) [arXiv:astro-ph/0609120];

T Biswas and A Notari, *JCAP* **0806**:021 (2008) [arXiv:astro-ph/0702555];

V Marra, E W Kolb, S Matarrese and A Riotto, *Phys. Rev.* **D76**, 123004 (2007) [arXiv:0708.3622];

V Marra, E W Kolb and S Matarrese, *Phys. Rev.* **D77**, 023003 (2008) [arXiv:0710.5505];

J Garcia-Bellido and T Haugboelle, *JCAP* **0804**:003 (2008) [arXiv:0802.1523];

I Jakacka and J Stelmach, *Class. Quantum Grav.* **18**, 2643 (2001) [arXiv:0802.2284];

C Yoo, T Kai and K Nakao, *Prog. Theor. Phys.* **120**, 937 (2008) [arXiv:0807.0932];

P Hunt and S Sarkar, arXiv:0807.4508 (2008);

R A Vanderveld, E E Flanagan and I Wasserman, *Phys. Rev.* **D78**, 083511 (2008) [arXiv:0808.1080];

J P Zibin, A Moss and D Scott, *Phys. Rev. Lett.* **101**, 251303 (2008) [arXiv:0809.3761];

T Clifton and J Zuntz, arXiv:0902.0726 (2009);

W Valkenburg, arXiv:0902.4698 (2009);

C. Clarkson and M. Regis, arXiv:1007.3443 [astro-ph.CO];

A. Moss, J. P. Zibin and D. Scott, arXiv:1007.3725 [astro-ph.CO];

S. Foreman, A. Moss, J. P. Zibin and D. Scott, arXiv:1009.0273 [astro-ph.CO].

[68] M E Araujo, W R Stoeger, R C Arcuri and M L Bedran, *Phys. Rev.* **D78**, 063513 (2008) [arXiv:0807.4193].

[69] R R Caldwell and A Stebbins, *Phys. Rev. Lett.* **100**, 191302 (2008) [arXiv:0711.3459];

J-P Uzan, C Clarkson and G F R Ellis, *Phys. Rev. Lett.* **100**, 191303

(2008) [arXiv:0801.0068];

K Bolejko and J S B Wyithe, *JCAP* **0902**:020 (2009) [arXiv:0807.2891];

C Quercellini, M Quartin and L Amendola, arXiv:0809.3675 (2008).

Chapter 5

Signals of cosmic magnetic fields from the cosmic microwave background radiation

T. R. Seshadri

Department of Physics and Astrophysics,
University of Delhi, Delhi 110 007, India.
E-mail: *sesh@iucaa.ernet.in*

Abstract: The Cosmic Microwave Background Radiation (CMBR) is the cleanest and most direct source of information about the physical processes in the universe at the recombination era. It also has the potential to give us indirect information about the very early universe. Depending on the ionization history of the universe, CMBR studies can help us understand physical processes in the post recombination era.

I start with the introduction to CMBR. I then present a brief discussion on causes of its anisotropy and polarization. Divergence-free velocity fields in baryons lead to characteristic signatures in the polarization (B-type polarization) of the CMBR. Normally such fields decay with cosmological expansion. However in the presence of tangled magnetic fields such divergence velocity fields can be sustained. I will discuss as to how, by studying the nature of polarization of the CMBR, and the non-Gaussianity of the temperature anisotropy, we can understand the strength and behavior of cosmic magnetic fields.

5.1 Introduction

Various techniques have been developed in recent years to investigate the nature of the universe. Among them, the analysis of the Cosmic Microwave Background Radiation has now established itself as the most important of these techniques. This is due to the the nature of interaction of photons with matter and gravity. One of the main reasons for this is that it is believed to have originated when the universe was about 300,000 years old. For comparison and to appreciate how early that epoch is one may note that the present age is about 15 billion years. Starting from such an early epoch, it reaches us almost completely preserving its nature. As a result through observations and a detailed understanding of the physics of CMBR, we can learn about the nature of the universe at the epoch of decoupling. As we will see, in a simple ideal model of the universe, the CMBR should be isotropic, unpolarized and have a black body spectrum. Processes at the surface of last scatter as well as between the epoch of decoupling and today, can introduce small deviations from the ideal behaviour. By assuming a robust model of the universe, and because of the fact that the deviations are small perturbations from the ideal behaviour, it is possible to extract more information about our universe and its evolution and constituents. In other words, CMBR has the record of information of the history of the universe. Further, there are various characteristics of the CMBR. First of all it can have different fluxes in different directions. This is referred to as anisotropy. Even if the fluxes were same, the polarization can be different in different directions. Also, the spectrum can show deviations from the black body nature. Since there are a variety of attributes to CMBR, these observations have the potential to break the degeneracy in estimation of cosmological parameters. All these features make the CMBR studies a potentially important probe to investigate the details of the universe. In this article, we will give a rather pedagogical exposition to the anisotropy in the temperature and polarization of CMBR and what can be the role and contribution of primordial magnetic fields to these features in CMBR.

5.2 Origin of CMBR

It was discovered by Penzias and Wilson [1], that there is a flux of radiation from all directions in the sky. Although this discovery was made by them in 1965, several researchers had already predicted the existence of such a radiation [2]. It was believed to be (and now shown observationally) that this radiation has a black body spectrum. It is called Microwave because the peak of the spectrum is in the microwave region. Further, it is called a background radiation because it is does not originate from a localized region but seems to be coming from all directions. While these features are observational, the reason for calling it Cosmic is interpretational. There are theories which do not consider this radiation to be cosmic. We will however, not be discussing these. We assume that this is indeed cosmic and hence is intricately intertwined with the cosmological model we use. Hence, it will be in place to discuss the the dynamics of the homogeneous universe and how it dictates the evolution of CMBR.

5.2.1 *Homogeneous universe*

The spacetime line element for a Minkowski spacetime is given by,

$$ds^2 = dt^2 - (dx^2 + dy^2 + dz^2) \tag{5.1}$$

The spatial distance is defined as the value of $\sqrt{-ds^2}$ at a constant time hypersurface. Similarly time interval is defined as $\sqrt{ds^2}$ at constant spatial coordinates.

The metric which describes cosmology should have a parameter characterizing the expansion of the universe. When we say that the universe is expanding, we mean that the distance between any two points in the universe increases with time. The line-element thus has a function of time multiplying the spacial part. Further, if the expansion is homogeneous and isotropic, the factor multiplying the spatial part is independent of position. Thus it depends only on time. We denote this function (called the scale factor) by $a(t)$. For an expanding universe this line-element turns out to be,

$$ds^2 = dt^2 - a^2(t)[dx^2 + dy^2 + dz^2] \tag{5.2}$$

which in spherical polar coordinates takes the form,

$$ds^2 = dt^2 - a^2(t)[dr^2 + r^2 d\theta^2 + r^2 \sin^2(\theta) d\phi^2] \tag{5.3}$$

For an expanding universe, $a(t)$ is an increasing function of time. This is not the only form of the line-element possible. There are other possibilities too for a homogeneous and isotropic universe. In general the line element is given by,

$$ds^2 = dt^2 - a^2(t) \left[\frac{dr^2}{1 - k/r^2} + r^2 d\theta^2 + r^2 \sin^2(\theta) d\phi^2 \right] \tag{5.4}$$

The constant $k = \pm 1, 0$. It will turn out that in certain situations it is more convenient to work with another time coordinate called the conformal time, η defined by,

$$d\eta = dt/a(t) \tag{5.5}$$

The line element in terms of these coordinates is given by,

$$ds^2 = a^2(\eta) \left[d\eta^2 - \frac{dr^2}{1 - k/r^2} - r^2 d\theta^2 - r^2 \sin^2(\theta) d\phi^2 \right] \tag{5.6}$$

(Strictly speaking we should denote the scale factor in terms η by some other function, say, $\tilde{a}(\eta)$. However, since it will be clear from the context, we can afford to denote it by the same symbol $a(\eta)$). Using conformal time has a big advantage. For photon trajectories, we have $ds^2 = 0$ and for any material particle with non-zero mass, $ds^2 > 0$. Clearly for a photon that travels radially (so that without loss of generality we may take $\theta = 0, \phi = 0$), we have $d\eta = \pm dr/\sqrt{1 - k/r^2}$. The evolution of the scale factor is dictated by the nature of the constituents of the universe. Using the time-time component of the Einstein equations we have,

$$\frac{\dot{a}^2}{a^2} + \frac{1}{a^2} = \frac{8\pi G\rho}{3} \tag{5.7}$$

The energy momentum tensor is divergenceless. This implies that the density and pressure follow the continuity equation,

$$\frac{d(\rho a^3)}{dt} + P\frac{d(a^3)}{dt} = 0. \tag{5.8}$$

If we compliment this by the equation of state that relates pressure to density

$$P = w\rho \tag{5.9}$$

we can arrive at the evolution of density with scale factor,

$$\rho \propto a^{-3(1+w)} \tag{5.10}$$

The evolution of the scale factor at any time will be governed by the total energy density in the universe. If a particular component is dominant, then time evolution of the scale factor will primarily be governed by that component. If we consider a $k = 0$ (i.e. spatially flat) the scale factor evolution with time is given by (from equations 5.8 and 5.10)

$$a \propto t^n, \tag{5.11}$$

where $n = \frac{2}{3}(1 + w)^{-1}$. For most kinds of matter, $0 \leq n < 1$.

For radiation the equation of state parameter, $w = 1/3$. With this value of w, we see from equation (5.10) that the radiation density, $\rho \propto a^{-4}$. This behavior has a simple interpretation. The number density of photons drops as the inverse of volume and hence, as a^{-3}. Further, due to cosmological redshift, the energy of every photon drops as a^{-1}. These two effects combine to make the radiation energy density evolve as a^{-4}.

With $w = 1/3$ for radiation, we see that the energy density evolves as $\rho \propto t^{-2}$. This simple calculation shows that the energy density of radiation (and hence its temperature) was higher in the past, and earlier the epoch we consider, higher is the density and radiation temperature.

The CMBR which we observe today is believed to be this relic radiation of the early universe. The spectrum of the CMBR is a black body spectrum. The energy density of the black body spectrum is proportional to the fourth power of temperature, i.e. $\rho \propto T^4$. Together with the fact that $\rho \propto a^{-4}$, we find that the temperature of radiation is inversely proportional to the scale factor, a. Since $a \propto t^n$, we see that the temperature of radiation in the past was higher. Further, the earlier the epoch we consider, higher is the temperature. Hence, it is expected that at a sufficiently early era, the temperature of the universe was high enough to ionize·the atoms and maintain them in the ionized state.

5.3   Origin of CMBR and the homogeneity of the universe

The nature of the CMBR is closely related to the thermal and ionization history of the universe. The baryonic matter in the universe consisted of about 75% hydrogen and 25% helium. The temperature of radiation is believed to have been high in the past and subsequently dropped with the cosmological expansion. The earlier the epoch higher was the temperature. Hence, as we study the nature of radiation at earlier times, we would reach an epoch when the temperature of radiation is just enough to sustain matter in the ionized form. At epochs prior to this the temperature of the radiation must have been high enough to keep matter ionized. The temperature needed to sustain matter in the ionized form depends on the ionization energy of matter. For hydrogen the ionization energy is 13.6 eV. From the time-temperature relation of the radiation this determines an epoch when the radiation temperature is sufficient to ionize the matter. The epoch when this happened is called the epoch of recombination. We denote this epoch by t_r and the corresponding scale factor by a_r. The temperature when this happens is about 3000K. Prior to this epoch, the temperature would have been even higher so that the matter could be sustained in the ionized form.

In order to calculate the degree of ionization at a given temperature one uses the Saha ionization formula [11],

$$\frac{X_e^2}{1 - X_e} = \frac{1}{n_e + n_H} \left[\left(\frac{m_e T}{2\pi} \right)^{3/2} e^{-(m_e + m_p - m_H)/T} \right], \qquad (5.12)$$

where, X_e is the free electron fraction and is equal to

$$X_e = \frac{n_e}{n_e + n_H} \tag{5.13}$$

where, n_e and n_H are the number densities of free electrons and hydrogen atoms, respectively, while m_e, m_p and m_H are the masses of electron, proton and hydrogen atom, respectively. At $kT \gg (m_e + m_p - m_H)c^2$, the degree of ionization X_e is almost unity. With cosmological expansion, the temperature drops. At a particular temperature called the ionization temperature, the degree of ionization drops rapidly to zero. Although this transition is not instantaneous, it is fairly abrupt. Since the temperature drops with time due to cosmological expansion, the matter in the universe makes a transition from plasma state to a state of neutral atoms. It is because of this sharp transition that we can identify a clear epoch of recombination.

Since photons couple strongly with charged particles, during the plasma phase the photons undergo significant scattering. Before the epoch of recombination, the photons undergo a series of steps of a random walk due to scattering from charged particles. Hence, any information encoded in the photons before the recombination epoch tends to get wiped out. On the other hand, after the epoch of recombination, the electrons and nuclei combine to form neutral atoms. Hence, the scattering of photons becomes negligible. Most of the photons can travel unhindered in a straight line. [We say most because in reality there is a small but non-zero degree of ionization ($X_e = 10^{-5}$) even after the recombination epoch.] In order to capture the basic essence of the process, the situation in $1 + 2$ dimensions is shown in Fig. 5.1. During the pre-recombination epoch, the photons keep getting scattered into random directions. After every scattering event, the photon moves freely along a light cone till it undergoes the next scattering at a different location (which is roughly one mean free path distance away) at a future time. After this scattering again the photon goes in a random direction but still along the light cone. The photon thus undergoes a Brownian motion. This process of alternate scattering and free flight continues till the epoch of recombination, t_r. In order to capture the essential physics, we can consider the recombination as an instantaneous process as already pointed out. The last scattering launches the photon as before in a random direction just as the earlier scattering events did. However, since there are no more free charged particles available after this epoch, the photon does not get re-scattered.

Consider a point P as shown in the figure at the epoch of recombination.

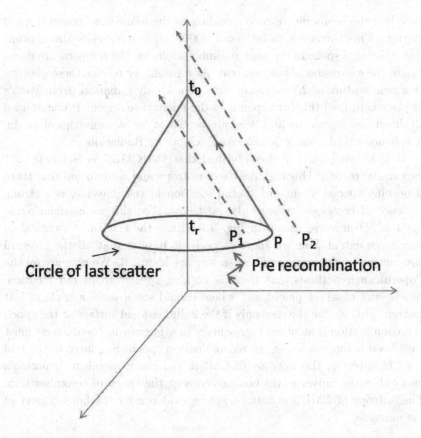

Fig. 5.1   Circle of Last Scatter in a $1+2$ dimensional universe. (Curr. Sc. **97** (2009) 858)

From point P there will be photons emitted in all directions. Photons from a point P_1 and directed towards the observer would have reached the observer's location and gone past at an earlier epoch. Similarly, a photon which got last scattered for the last time at a point P_2 has not reached the observer till today. The locus of the points from where the photons would have reached the observer at t_0 form a circle. Thus the observer perceives an apparent circle from where the photons are reaching him/her. This can be called the 'Circle of Last Scatter'. For the case of $1+3$ dimensional universe that we inhabit, instead of the Circle of Last Scatter we will have a Spherical Surface of Last Scatter with the observer at its centre. The photons when they started out from this surface must have had the temperature corresponding to the ionization temperature of matter. This turns out to be about 3000K. However, while the photons travel through space in the post recombination era, the temperature drops due to the expansion of the universe. As a result, we receive these photons at a temperature of 2.73K today. For a blackbody radiation (from Wein's displacement law) this corresponds to the microwave region. It comes from all directions around us and it is interpreted to be of cosmological origin. It is hence called Cosmic Microwave Background Radiation.

It is by now fairly well-established that the CMBR is isotropic to 1 part in 10^5 to 10^6. This high degree of isotropy and assumption that there is nothing special about our spatial location in the universe, is a strong evidence of homogeneity as explained below. For this we consider again the $1 + 2D$ universe shown in Fig. 5.1. Since the CMBR is identical to very large extent from all the directions, it implies that all the physical parameters on the surface of last scatter are identical. We now invoke the Copernican hypothesis that there is nothing special about our location. Hence, any observer placed any where should see a similar circle of last scatter. This can be the case only if the 2 dimensional surface at the epoch of recombination is identical everywhere. In other words, the universe must have been homogeneous at the recombination epoch. Similarly, in the real $1 + 3D$ universe, the isotropy of CMBR and the Copernican hypothesis imply that the universe was homogeneous at the epoch of recombination. The isotropy of CMBR is hence a strong evidence for the homogeneity of the universe.

5.4   Finer features of the CMBR: A brief introduction

The CMBR as we saw in the last section is predominantly isotropic, unpolarized and has a blackbody spectrum. However, there are small deviations that are expected from different processes in the evolution of the universe. Some of these deviations have already been observed from several observational missions and it is hoped that in the coming years many more will be observed. There are broadly three different features that are embedded in an other wise isotropic, unpolarized and blackbody CMBR.

(1) Angular dependence: Intensity of radiation is almost isotropic but can have small dependence in direction.
(2) Polarization: Degree of polarization can be slightly different in

different directions.

(3) Spectrum: The radiation is almost Planckian but can have slight deviations from it.

In this article, aspects relating to polarization are discussed. For a detailed discussion on different aspects of CMBR, a number of reviews [4, 5] and textbooks [6, 7, 8, 9] are available.

5.4.1 *Temperature anisotropy*

Let us once again start with the $1 + 2D$ spacetime as an illustrative example and extend it then to the real $1 + 3D$ case. We had noted that the photons we receive are from a circle of last scatter which is located at the epoch of recombination. We had also argued that if the universe were homogeneous at that time, the nature of the photons received from all the points on the circle would have been similar. The observer would feel that she/he is at the centre of a circle of uniform temperature from which the incoming radiation originates. However, if the points on the CLS are not identical, the signals coming from different directions are different. Hence, the inhomogeneity at the epoch of recombination would show up as anisotropy. Similarly, when there is inhomogeneity in the real situation of $1 + 3D$ universe, the observer would feel that she/he is at the centre of a hypothetical sphere (the spherical surface of last scatter). If there were inhomogeneities at the epoch of recombination, then the signals received from different directions on the spherical surface of last scatter will be different and hence, the inhomogeneity would translate as the anisotropy in the CMBR.

This inhomogeneity could originate due to different reasons. The gravitational potential could be different at different portions of the circle. This can arise if there are inhomogeneities in the dark matter distribution. Further the baryon density can have inhomogeneities. Also the baryon velocity fields could be different in different regions. From Helmholtz theorem, any vector field can be split as a sum of the gradient of a scalar (curl free) and curl of a vector (divergence free) parts. Velocity fields arising due to gravitational potential belong to the first category, while those arising from Lorentz force due to primordial magnetic fields belong to the latter category. It would often prove useful to express the temperature not in the terms of angles but in terms of the corresponding Fourier modes. It will be instructive to once again look at the corresponding $1 + 2D$ situation.

Let the temperature of radiation from the circle of last scatter be expressed as $T(\theta)$. We may define an average temperature as

$$T_0 = \frac{1}{2\pi} \int_0^{2\pi} d\theta \ T(\theta) \tag{5.14}$$

We may then define a fractional deviation of temperature from this mean value as

$$T(\theta) = T_0(1 + \epsilon(\theta)) \tag{5.15}$$

Since this function is defined on a circle, it can be expanded as a Fourier series,

$$\epsilon(\theta) = \sum a_n \cos(n\theta) + b_n \sin(n\theta), \tag{5.16}$$

and it is more convenient to work in terms of these Fourier coefficients, a_n's and b_n's.

In the $1 + 3$ dimensional universe, the temperature in any direction is denoted by $T(\theta, \phi)$. We define a mean temperature,

$$T_0 = \frac{1}{4\pi} \int_{\theta=0}^{\pi} \int_{\phi=0}^{2\pi} \sin(\theta) \ d\theta \ d\phi \ T(\theta, \phi) \tag{5.17}$$

The temperature in any direction can be expressed as,,

$$T(\theta, \phi) = T_0[1 + \Theta(\theta, \phi)] \tag{5.18}$$

where, $\Theta(\theta, \phi)$ is the fractional perturbation in the temperature, $\Delta T/T_0$. As in the case of $1 + 2$ dimension it will prove more useful to expand Θ in terms of a set of basis functions. In this case these are the spherical harmonics $Y_{lm}(\theta, \phi) = Y_{lm}(\hat{n})$ which form a set of complete basis functions defined on the celestial sphere.

$$\Theta_{lm} = \int d\Omega \ Y_{lm}^*(\hat{n}) \ \Theta(\hat{n}) \tag{5.19}$$

The anisotropy is characterized by the angular two-point correlation function of $\Theta(\hat{n})$. Explicitly, this is defined as $\langle \Theta(\hat{n})\Theta(\hat{m}) \rangle$ which, in general, can be expected to be a function of the directions $\hat{n}$ and $\hat{m}$. However, in a Friedmann universe, the correlation proves to be a function only of α, viz. the angle between the directions $\hat{n}$ and $\hat{m}$, i.e. one has $C(\alpha) = \langle \Theta(\hat{n})\Theta(\hat{m}) \rangle$. The reason for this is that due to isotropy, what matters is not which direction the instrument is facing but only on the angle between the two directions. If the temperature anisotropy has a Gaussian behavior, the anisotropy should be completely describable by the power spectrum

$$C_l \delta_{ll'} \delta_{mm'} = \langle \Theta_{l'm'}^* \Theta_{lm} \rangle \tag{5.20}$$

The fact that the power spectrum C_l is independent of m has just to do with the fact that the quantity $C(\alpha)$ depends only on the angle between $\hat{n}$ and $\hat{m}$. One may look at it in the following way. Since no particular direction is special, we can take the direction $\hat{n}$ to be the z direction. Thus α becomes the polar angle. Further, since the direction $\hat{m}$ is only constrained to make an angle α with $\hat{n}$ but is otherwise arbitrary, the azimuthal angle has no role.

5.5 Origin of temperature anisotropy in the CMBR

The Boltzmann equation describing the evolution of the distribution function is given by

$$\frac{df}{dt} = \frac{\partial f}{\partial t} + \frac{dx^i}{dt}\frac{\partial f}{\partial x^i} + \frac{dp_j}{dt}\frac{\partial f}{\partial p_j} = C[f] \qquad (5.21)$$

For CMBR photons, the collision term on the right hand side is mainly governed by the Compton scattering process between electron and photons. From the moments of the Boltzmann equation we can arrive at the equations for the evolution of excess temperature of the CMBR. With these two in equation (5.21) we can derive the evolution equation of the radiation temperature. The equation governing temperature anisotropy turns out to be,

$$\dot{\Theta} + ik\Theta = -\dot{\Phi} - ik\mu\Psi - \dot{\tau}\left[\Theta_0 - \Theta + \mu v_b - \frac{1}{2}P_2(\mu)\Pi\right] \qquad (5.22)$$

(For a detailed discussion on this the reader is referred to [11] and [10]). Here, Θ is the Fourier transform of the temperature anisotropy and v_b is the velocity field of baryons in Fourier space. The term Π is the sum of quadrupole terms of temperature anisotropy, polarization anisotropy and the polarization monopole term. Φ and Ψ are perturbations in the metric and are related to the gravitational potential perturbations. These perturbations are dominant typically over angular scales of more than $1°$. The factor, τ is the optical depth. The temperature anisotropy over the largest angular scales arise from the perturbations in the gravitational potential on the surface of last scatter. This is referred to as the Sachs-Wolfe effect.

As the photon travels from the surface of last scatter, it passes through structures which are still in the process of collapse. The photon thus experiences a deepening time dependent gravitational potential. This leads to the integrated Sachs-Wolfe effect. The baryon and the photon fluid are tightly coupled to each other. It behaves like a single fluid. Perturbations in this fluid undergo acoustic oscillations. These oscillations at the SLS manifest themselves as crests and troughs in the anisotropy power spectrum. The angular size for a physical length scale depends on the background geometry. However, the correspondence between the angular size and the physical length scale is not one-to-one. Due to projection effect, a range of physical length-scales will correspond to the same angular scale. The contribution to the angular power spectrum for a certain l comes mainly from $l = k(\eta_0 - \eta_r)$. Here η_0 and η_r are the conformal times corresponding to todays epoch and the epoch of recombination, respectively. Because of the projection effect, however, this value of l is not the only contribution but only the predominant one.

The structure of the angular power-spectrum contains information about parameters describing the cosmology. This angular power-spectrum consists of a series of peaks. The values of l at which the peaks occur are $l = n\pi \frac{\eta_0 - \eta_r}{r_s(\eta_r)}$. Here $r_s(\eta_r)$ is the comoving sound horizon at the epoch of recombination. The curvature of the universe has a bearing on the location of the first peak. Due to the non-zero value of the baryon density, the odd peaks tend to have a larger value than the even ones.

5.6   Characterizing the nature of CMBR polarization anisotropy

Several features and processes in the universe can lead to a polarization of the CMBR. Although polarization is a much weaker signal in general, as compared to temperature anisotropy, with the development of newer experimental techniques one has started seeing signals of polarization anisotropy, and the situation is expected to improve with the satellite borne experiments that are planned. Hence, the study of CMBR polarization is gaining increasing importance and relevance more than ever before. Here we review these characteristics in brief. For details there are very good reviews and textbooks available [11, 12, 8, 9, 3].

Polarization is described in terms of the Stokes parameters, I, Q, U and V. Here, I is the total intensity. Its information is encoded in the temperature anisotropy. As we will see in the next section, polarization in CMBR is caused due to Thomson scattering of anisotropic radiation scattered from free electrons. Such a process can lead to a non-zero degree of plane polarization but cannot produce circularly polarized light. The Stokes parameter

V is a measure of circular polarization. Hence, in polarization processes in the context of cosmology, this parameter is taken to be zero. Thus we may assume that to describe polarization in the context of cosmology, we could work with the two Stokes parameters, Q and U. The Stokes parameter Q measures the polarization in the vertical and horizontal directions. U measures the polarization at $\pm 45^o$ with the horizontal. It is immediately clear that a pure Q-type polarization goes to pure U-type just by the rotation of the coordinate system. Ideally one would like to have a description which does not depend on the coordinate system. One might naively think that this is possible if we define a particular direction as universal. This is true in a two dimensional Euclidean surface but not on the celestial sphere as it is non-Euclidean. This can be seen through the following example. If we have a horizontal plane described by a 2-dimensional Euclidean Geometry, then one can define a universal reference axis and specify orientations of straight lines with respect to this. The specification of this direction is unique no matter where the straight line is located.

The situation is not so simple when we are describing orientations on the surface of a sphere. Two small line segments at well separated locations on a spherical surface, both of which are pointing north-south will have two different orientations. The situation is similar in the case of polarization. The polarization directions are specified on the celestial sphere which, obviously, is non-Euclidean. To measure/compute polarization anisotropy, we need to correlate polarization orientation at two different locations. Due to the complication arising out of non-Euclidean geometry, this exercise is non-trivial if we simply use the Stokes parameters Q and U. If we were to correlate the polarization orientation in two directions $\hat{n}$ and $\hat{m}$ that subtend a small angle with each other, then of course one can make a flat sky approximation over a small region on the celestial sphere. (By small angle we mean, $\cos^{-1}(\hat{n}.\hat{m}) \ll 2\pi$.) Since in this small region the deviation from Euclidean nature is small, we can follow the same procedure as in the case of a Euclidean 2 dimensional plane. We can define a direction in the flat patch on the celestial sphere. The polarization orientation correlation at two different points in this nearly flat patch will not have any significant artifact due to the non-Euclidean nature. However, if we need to measure or describe the correlation at two points that are well separated, we cannot make a flat sky approximation.

We know that a spinor which is invariant under a rotation of 4π radians is described by a spin-1/2 field. Similarly vectors which are invariant under a rotation by 2π are described by a spin-1 field. Polarization is invariant under a rotation by π and is described by a spin-2 field.

Following reference [8], we define the dimensionless Stokes parameters

$$\mathcal{Q} = \frac{Q}{4I} \quad \mathcal{U} = \frac{U}{4I} \tag{5.23}$$

Under rotation by an angle ψ these quantities transform as,

$$Q' \pm iU' = \exp^{\pm i2\psi}(Q \pm iU) \tag{5.24}$$

$$I' = I \tag{5.25}$$

$$V' = V \tag{5.26}$$

and hence,

$$\mathcal{Q}' \pm i\mathcal{U}' = \exp^{\pm i2\psi}(\mathcal{Q} \pm i\mathcal{U}) \tag{5.27}$$

Just as we expanded the temperature on the celestial sphere in terms of the spherical harmonics, a spin-2 field can be expanded on a spherical surface in terms of the spin weighted spherical harmonics, $_{\pm 2}Y_{lm}$.

$$(\mathcal{Q} \pm i\mathcal{U})(\hat{n}) = \sum_{l=2}^{\infty} \sum_{m=-l}^{l} a_{lm}^{\pm 2} \,_{\pm 2}Y_{lm} \tag{5.28}$$

$$= \sum_{l=2}^{\infty} \sum_{m=-l}^{l} (e_{lm} \pm ib_{lm}) \,_{\pm 2}Y_{lm} \tag{5.29}$$

Under parity transformation, $a_{lm}^{+2} \longleftrightarrow a_{lm}^{-2}$, $e_{lm} \longleftrightarrow e_{lm}$ and $b_{lm} \longleftrightarrow -b_{lm}$.

Our aim is to describe polarization anisotropy in terms of variables which like temperature anisotropy are invariant under rotation. Directly in terms of Stokes parameters, it is not possible as they are components of a spin-2 field. However, one can construct spin-0 fields by the following procedure. One can define spin raising (β) and lowering (β^*) operators. Operating this twice on spin-2 fields one can construct spin 0 fields which have this invariance property. The spin raising and lowering operators are similar to the raising and lowering operators one comes across in the context of angular momentum in quantum mechanics.

The spin 0 fields thus constructed are very much like temperature anisotropy and can be expanded in terms of the usual spherical harmonics. To this end one constructs these operators and are defined in terms of their action on the spin weighted spherical harmonics.

$$\beta^2(_{-2}Y_{lm}) = \left(\sqrt{(l+2)!/(l-2)!}\right) Y_{lm} \tag{5.30}$$

$$\beta^{*2}(_{+2}Y_{lm}) = \left(\sqrt{(l+2)!/(l-2)!}\right) Y_{lm} \tag{5.31}$$

We can then define polarization parameters denoted by $\mathcal{E}$ and $\mathcal{B}$ as,

$$\mathcal{E} = \frac{1}{2} \left[\not{\partial}^2 + \not{\partial}^{*2} \right] \mathcal{Q} \tag{5.32}$$

$$\mathcal{B} = -i\frac{1}{2} \left[\not{\partial}^2 - \not{\partial}^{*2} \right] \mathcal{U} \tag{5.33}$$

Using these raising and lowering operators on both sides of equation 5.28 we have,

$$\mathcal{E}(\hat{n}) = \sum_{l=2}^{\infty} \sum_{m=-l}^{l} e_{lm} \ Y_{lm}(\hat{n}) \tag{5.34}$$

$$\mathcal{B}(\hat{n}) = \sum_{l=2}^{\infty} \sum_{m=-l}^{l} b_{lm} \ Y_{lm}(\hat{n}) \tag{5.35}$$

These two variables are the ones which describe polarization while at the same time are invariant under rotation. We have already mentioned the transformation properties of e_{lm} and b_{lm} under parity transformation. Using this we find from the above expression that $\mathcal{B}$ has odd parity and $\mathcal{E}$ has even parity. Due to this property the former is called magnetic type and the latter the electric type polarization.

The electric and magnetic type polarization anisotropy is measured in terms of the power spectrum of the respective quantities. As in the case of temperature anisotropy, we define the power spectrum in electric and magnetic type polarization anisotropy as,

$$C_l^{\mathcal{E}} = \langle | \ e_{lm} \ |^2 \rangle \tag{5.36}$$

$$C_l^{\mathcal{B}} = \langle | \ b_{lm} \ |^2 \rangle \tag{5.37}$$

We can also define the cross-correlation $C_l^{\Theta\mathcal{E}}$ as,

$$C_l^{\Theta\mathcal{E}} = \langle \Theta_{lm} e_{lm} \rangle \tag{5.38}$$

The cross-correlation, $C_l^{\mathcal{E}\mathcal{B}}$ should be zero because $\mathcal{E}$ and $\mathcal{B}$ are of opposite parity. Similarly, $C_l^{\Theta\mathcal{B}}$ should also be zero. However, if there were parity violating processes then these would also be non-zero. Thus measurement of $C_l^{\mathcal{E}\mathcal{B}}$ and $C_l^{\Theta\mathcal{B}}$ can put bounds on parity violating processes.

5.7 Origin of CMBR polarization anisotropy

The possibility of a non-zero polarization anisotropy was suggested in 1968 [13]. The basic mechanism is illustrated in Fig 5.2. Light is incident from

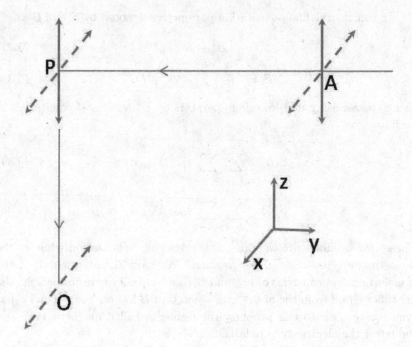

Fig. 5.2   Polarization from Thomson Scattering (Curr. Sc. **97** (2009) 858)

point A to the point P where there is a free electron. Let AP be parallel to the y-axis. Point P is vertically above a point O where the observer is located. OP direction is defined as the z-axis. For the incident light, the direction of electric field oscillations are in a plane parallel to the x-z plane. Hence, when this light disturbs the electron at P, it will oscillate only in the x-z plane i.e.. perpendicular to the y-axis. This oscillation will lead to radiation. Consider the ray of light traveling from P to O. Since the electron oscillation does not have a component along the y-axis. the electric field vector does not have a component in that direction. The oscillations along the the z-axis do not contribute to radiation. That implies that the radiation along PO has electric oscillations which are parallel only to the x-axis. Hence we find that although the incident light is unpolarized, the scattered light in the PO direction is plane polarized. If at sunrise or sunset, we observe the light from the zenith through a polarizer, we will find that the light has a non zero degree of polarization. The free electrons in the ionosphere, scatter the light coming from the horizon. When we look at the zenith at this time, the direction of scattered light that reaches the observer from the zenith, is perpendicular to the direction of the incoming light. If the incident light was from all direction with equal intensity, It is easy to see that the scattered light along PO is unpolarized. This implies that the incident light needs to to be anisotropic. It turns out that if the incident light has dipole anisotropy, this is still insufficient to produce polarization after Thomson Scattering. In order to produce polarization we need light with a quadrupole component of anisotropy to be Thomson scattered off free charged particles.

The universe is believed to have undergone re-ionization in the post recombination era. With a quadrupole component in the anisotropy of the CMBR this is one possible scenario for the origin of polarization. Another situation that can cause polarization is the finite thickness of the surface of last scattering. The electron that is responsible for the last scattering of a photon sees a quadrupole anisotropy from the last but one scattering. Thus even without re-ionization one does expect a polarization signal in the CMBR. We had earlier discussed the electric and magnetic type polarization. Scalar modes (arising due to density and potential perturbations) lead only to the electric type perturbations. Cosmic magnetic fields produce solenoidal velocity fields that lead to vector type perturbations. These perturbations produce magnetic type polarization [17, 14, 15, 16]. Models of inflation produce tensor perturbations in addition to scalar ones. The magnetic type polarization can also be produced by such tensorial perturbations.

5.8   Cosmic magnetic fields

We still do not know how magnetic fields in our universe originated. Turbulent dynamo action on small seed magnetic fields is one possible mechanism for the origin of these fields [18, 19]. This mechanism has strong constraints. A different line of approach is that magnetic fields on cosmological scales could be the remnants of primordial fields. These primordial fields could be generated in the early universe. Evolution of large-scale structures might have been influenced by a primordial field with a strength 10^{-9} G at the present epoch and tangled on galactic scales.

At this point it should be pointed out that we are considering tangled magnetic fields and not a uniform one. Since magnetic field is a vector, if we have a uniform magnetic field, it will explicitly violate the isotropy in the universe. Alternatively, limits on the uniform component of the magnetic field can be placed from the observed anisotropy in the CMBR. We would however, point out that we are not considering such fields in this article.

The fields we consider are those that are tangled over some length scale so that there is no violation of isotropy on this count over the scales on which these tangles are present. The consequences of such fields have been studied in literature [14, 20, 21, 15, 22, 23]. For a detailed numerical investigation refer to [16]. To describe such fields, we need to specify its statistical properties. We assume that the magnetic field is a Gaussian random field. In other words, it can be specified purely from the two-point correlation function.

The magnetic field induces velocity fields in baryons due to Lorentz force. If this velocity were large, then that would in turn produce magnetic field which will change the initial magnetic field. We assume that this does not happen. In other words, the induced velocity field is sufficiently small so as not to modify the original magnetic field. This feature ensures that the magnetic field simply redshifts with the expansion as,

$$\mathbf{B}(\mathbf{x}, t) = \mathbf{b}_0(\mathbf{x})/a^2(t) \tag{5.39}$$

The Lorentz force that drives the velocity field is thus,

$$\mathbf{F}_L = (\nabla \times \mathbf{b}_0) \times \mathbf{b}_0/(4\pi a^5) \tag{5.40}$$

Magnetic stresses also lead to metric perturbations. However, we do not consider these effects in this article.

For a statistical description we also need to specify its spectrum, $M(k)$. We define,

$$\langle b_i(\mathbf{k}) b_j(\mathbf{q}) \rangle = \delta_{\mathbf{k},\mathbf{q}} P_{ij}(\mathbf{k}) M(k) \tag{5.41}$$

where $\delta_{\mathbf{k},\mathbf{q}}$ is the Kronecker delta function. This gives,

$$\langle \mathbf{b}_0^2 \rangle = 2 \int \frac{dk}{k} \Delta_b^2(k) \tag{5.42}$$

where, $\Delta_b^2(k) = k^3 M(k)/2\pi^2$. This is the power per unit logarithmic interval in $k-$space. It is further convenient to define a dimensionless spectrum $h(k) = \Delta_b^2(k)/(B_0^2/2)$, where, B_0 is the fiducial constant magnetic field.

5.9 Polarization in CMBR due to magnetic fields

When light is incident on charged particles, the scattered light can be polarized if the incident light is anisotropic. Further, the nature of anisotropy should be such that it has a quadrupole component. In the context of cosmology, the quadrupole anisotropy of the CMBR could undergo Thomson scattering from charged particles in the universe and hence pickup a non-zero degree of polarization. The evolution equations for the lth multipole moments, Θ_l, E_l and B_l, of the temperature anisotropy $(\Delta T/T)$, the electric (E-) type and the odd parity, magnetic (B-) type polarization anisotropies, respectively, for vector perturbations, have been derived in detail by Hu & White [24] (see also [14, 23]). For vector perturbations, the B-type contribution dominates the polarization anisotropy ([24]). The details of its calculations and a summary of the results for the E-type contribution and the T-E cross correlations are given here. Polarization is sourced by the quadrupole anisotropy term given by $P(k, \eta) = [\Theta_2 - \sqrt{6}E_2]/10$, where k is the co-moving wave number and η is the conformal time. In the tight-coupling approximation, $(kL_\gamma(\eta) \ll 1$, where $L_\gamma(\eta)$ is the co-moving, photon mean free path) P can be analytically estimated. To leading order in this approximation, the quadrupole component is zero, and there is a non-zero dipole given by, $\Theta_1 = v_B$, where $v_B(k, \eta)$ is the magnitude of the rotational component of the fluid velocity v_i^B, in Fourier space. To the next order the quadrupole is non-zero. It is generated from the dipole at the 'last but one' scattering of the CMBR. Using Eq. (60), (63) and (64) of [24], we find that the quadrupole is given by, $\Theta_2 = -4E_2/\sqrt{6} = 4kL_\gamma v_B/(3\sqrt{3})$. Thus, $P = \Theta_2/4 = kL_\gamma v_B/(3\sqrt{3})$. Using this in Eq. (77) and (56) of [24] B_l is estimated and the angular power spectra C_l^{BB} arising from B-type polarization anisotropy ([14]),

$$
C_l^{BB} = 4\pi \frac{(l-1)(l+2)}{l(l+1)} \int_0^\infty \frac{k^2 dk}{2\pi^2} \frac{l(l+1)}{2}
$$

$$
\times <| \int_0^{\eta_0} d\eta g(\eta_0, \eta) (\frac{kL_\gamma(\eta)}{3}) v_B(k, \eta)
$$

$$
\times \frac{j_l(k(\eta_0 - \eta))}{k(\eta_0 - \eta)} |^2 > . \tag{5.43}
$$

Here $j_l(z)$ is the spherical Bessel function of order l, and η_0 the present value of the conformal time, η. The 'visibility function', $g(\eta_0, \eta)$, determines the probability that a photon reaches us at epoch η_0 if it was last scattered at an epoch in the interval η and $\eta + d\eta$. We assume the universe to be spatially flat. The total matter density with Ω_m and a non-zero cosmological constant density with $\Omega_\Lambda = 1 - \Omega_m$ is supposed to be the dominant components of the constituents of the universe today.

Let us briefly discuss the nature of the visibility function. Before the epoch of recombination, the matter in the universe was highly ionized (degree of ionization ~ 1). Now if we consider a CMBR photon, its last scattering event could not have been much earlier than the epoch of recombination. That is because if it was not the case, it could not have survived without undergoing more scattering as the matter is ionized. Hence, a scattering event much before recombination could not have been the last scattering event. Recall that the visibility function $g(\eta_0, \eta)d\eta$, is the probability that a photon that reaches us at the present epoch (η_0) was last scattered at an epoch in the interval η and $\eta + d\eta$. Thus the value of $g(\eta_0, \eta)$ for $\eta \ll \eta_{rec}$ is very small. Similarly for $\eta \gg \eta_{rec}$ again the visibility function is small as there are few ionized particles to scatter photons. Further as the visibility function has the interpretation of probability, its integral should come out to be unity. Since the contribution to visibility function is small, both much before and much after the recombination, the dominant contribution should come from the times around recombination. We approximate the visibility function as a Gaussian. $g(\eta_0, \eta) = (2\pi\sigma^2)^{-1/2} \exp[-(\eta - \eta_{rec})^2/(2\sigma^2)]$, where, η_{rec} is the conformal epoch of "last scattering" and σ measures the width of the LSS. The conformal time and the redshift is related by, $\eta = 6000h^{-1}((a+a_{eq})^{1/2} - a_{eq}^{1/2})/\Omega_m^{1/2}$, (We have here assumed the universe to be spatially flat [24].) To estimate these, we use the WMAP results [25], that the redshift of LSS $z_{rec} = 1089$ and its thickness (FWHM) $\Delta z = 194$. The expansion factor $a = (1+z)^{-1}$ and $a_{eq} = 4.17 \times 10^{-5}(\Omega_m h^2)^{-1}$ (h is the Hubble constant in units of 100 km s^{-1} Mpc^{-1}). For the Λ-dominated model suggested by WMAP results, with $\Omega_m = 0.27$, $\Omega_\Lambda = 0.73$, we have $\eta_{rec} = 201.4h^{-1}$ Mpc and $\sigma = 11.5h^{-1}$ Mpc. For an $\Omega_m = 1$ model, with the baryon density parameter, $\Omega_b = 0.0224h^{-2}$, we use the expressions given in [26] to estimate $\eta_{rec} = 131.0h^{-1}$ Mpc, and $\sigma = 8.3h^{-1}$ Mpc. We use these numbers in the numerical estimates below.

To calculate C_l^{BB}, we need to calculate the rotational velocity induced by magnetic inhomogeneities. We have already mentioned that the magnetic field is assumed to be initially a Gaussian random field and that on galactic scales and above, the induced velocity is generally so small that it does not lead to any appreciable distortion of the initial field [27, 28]. So, the magnetic field simply redshifts away as $\mathbf{B}(\mathbf{x}, t) = \mathbf{b}_0(\mathbf{x})/a^2$. The Lorentz force associated with the tangled field is then $\mathbf{F}_L = (\nabla \times \mathbf{b}_0) \times \mathbf{b}_0/(4\pi a^5)$, which drives the fluid and creates rotational velocity perturbations. These can be estimated as in [20] or [14], by using the Euler equation for the baryons. On scales larger than the photon mean-free-path at decoupling, the viscous effect due to photons can be treated in the diffusion approximation. The equation of motion for the fluid then takes the form,

$$\left(\frac{4}{3}\rho_\gamma + \rho_b\right)\frac{\partial v_i^B}{\partial t} + \left[\frac{\rho_b}{a}\frac{da}{dt} + \frac{k^2\eta}{a^2}\right]v_i^B = \frac{P_{ij}F_j}{4\pi a^5}. \tag{5.44}$$

Here, ρ_γ is the photon density, ρ_b the baryon density, and $\eta = (4/15)\rho_\gamma l_\gamma$ the shear viscosity coefficient associated with the damping due to photons. The mean-free-path of the photons is $l_\gamma = (n_e\sigma_T)^{-1} \equiv L_\gamma a(t)$, where n_e is the electron density and σ_T the Thomson cross-section. We have also ignored here a metric perturbation term which is subdominant at large l (cf. [14]). For $z_{rec} \sim 1089$, we get $L_\gamma(\eta_{rec}) \sim 1.83(\Omega_b h^2/0.0224)^{-1}$ Mpc. We have defined the Fourier transforms of the magnetic field, by $b_0(x) = \sum_k b(k)\exp(ik.x)$ and $F(k) = \sum_p[b(k+p).b^{rec}(p)]p - [k.b^{rec}(p)]b(k+p)$. The projection tensor, $P_{ij}(k) = [\delta_{ij} - k_i k_j/k^2]$ projects F onto its transverse components perpendicular to k.

The comoving Silk damping scale at recombination, $L_S = k_S^{-1} \sim 10$ Mpc, gives a natural length scale on the basis of which we can divide length scales in two types. For scales that are small compared to the Silk damping scale ($kL_S \gg 1$) the radiative viscosity is important. On the other hand, for scales much larger than the Silk damping scale, ($kL_S \ll 1$) radiative viscosity is negligible. For the latter case, the Lorentz force will dominate over photon viscosity. Assuming negligible initial rotational velocity perturbation, we can integrate the baryon Euler equation to get $v_i^B = G_i D$, where $G_i = 3P_{ij}F_j/[16\pi\rho_0]$ and $D = \eta/(1 + S_{rec})$ (see [20] or paper I). Here ρ_0 is the present-day value of ρ_γ, and $S_{rec} = (3\rho_b/4\rho_\gamma)(\eta_{rec}) \sim 0.59f_b$. For $kL_s \gg 1$, we can neglect the inertial term and use the terminal-velocity approximation, in the Euler equation, to balance the Lorentz force by friction. This gives $v_i^B = G_i(k)D$, but with D now given by $D = (5/k^2 L_\gamma)$, on scales where diffusion damping operates. The transition Silk scale can also be estimated by equating v_i^B in the two cases, to give $k_S \sim [5(1 + S_{rec})/(\eta L_\gamma(\eta))]^{1/2}$.

In order to compute C_l^{BB}s spectrum of the tangled magnetic field, denoted by $M(k)$, needs to be specified. With this aim we define, $\langle b_i(k)b_j(q)\rangle = \delta_{k,q}P_{ij}(k)M(k)$, where $\delta_{k,q}$ is the Kronecker delta which is non-zero only for $k = q$ This implies $< b_0^2 >= 2\int(dk/k)\Delta_b^2(k)$, where $\Delta_b^2(k) = k^3 M(k)/(2\pi^2)$ is the power in magnetic tangles per logarithmic interval in k space. The ensemble average $< |v_B|^2 >$, and hence the C_l^{BB}s, can be computed in terms of the magnetic spectrum $M(k)$. It is convenient to define a dimensionless spectrum, $m(k) = \Delta_b^2(k)/(B_0^2/2)$, where B_0 is a fiducial constant magnetic field. The Alfvén velocity, V_A, for this fiducial field is,

$$V_A = \frac{B_0}{(16\pi\rho_0/3)^{1/2}} \approx 3.8 \times 10^{-4} B_{-9}, \tag{5.45}$$

where $B_{-9} \equiv (B_0/10^{-9} \text{Gauss})$. We consider power-law magnetic spectra, $M(k) = Ak^n$ with a cut-off at $k = k_c$, where k_c is the Alfvén-wave damping length-scale ([28, 27]). The constant A is fixed by demanding that the smoothed field strength over a "galactic" scale, $k_G = 1h\text{Mpc}^{-1}$, (using a sharp k-space filter) is B_0, giving a dimensionless spectrum for $n > -3$ of

$$m(k) = (n+3)(k/k_G)^{3+n}. \tag{5.46}$$

The nature of the visibility function implies that the dominant contributions to the integral over η in Eq. (5.43) come from a range σ around the epoch $\eta = \eta_{rec}$. Further, $j_l(k(\eta_0 - \eta))$ ensures that those values of (k, η) dominate the integrand which have $k(\eta_0 - \eta) \sim l$. Thus, following the arguments detailed in [14] and [20], for $k\sigma \ll 1$ we get the analytical estimate,

$$l(l+1)C_l^{BB}/(2\pi) \approx (kL_\gamma(\eta rec)/3)^2(\pi/4)\Delta_v^2(k, \eta_{rec})|_{k=l/R_{rec}}. \tag{5.47}$$

Here, $\Delta_v^2 = k^3 < |v_B(k, \eta_{rec})|^2 > /(2\pi^2)$ is the power per unit logarithmic interval of k, in the *net* rotational velocity perturbation, and $R_{rec} = \eta_0 - \eta_{rec}$. In the other limit, $k\sigma \gg 1$, we get $l(l+1)C_l^{BB}/(2\pi) \approx (kL_\gamma(\eta rec)/3)^2(\sqrt{\pi}/4)(\Delta_v^2(k, \eta_{rec})/(k\sigma)|_{k=l/R_{rec}}$. At small wavelengths, C_l^{BB} is suppressed by a $1/k\sigma$ factor due to the finite thickness of the Surface of Last Scatter (SLS). Further in both cases, the polarization anisotropy, $\Delta T_P^{BB}(l) \approx (kL_\gamma(\eta_*)/3) \times \Delta T(l)$, where, $\Delta T(l)$ is the temperature anisotropy computed in [20]. Approximate analytic estimates for the CMBR polarization anisotropy induced by tangled magnetic fields can be derived using the above results. We define the quantity $\Delta T_P^{BB}(l) \equiv [l(l+1)C_l^{BB}/2\pi]^{1/2}T_0$, where $T_0 = 2.728$ K is the CMBR temperature. On large scales, such that $kL_s < 1$ and $k\sigma < 1$, the resulting CMBR anisotropy is (see [14])

$$\Delta T_P^{BB}(l) = T_0(\frac{\pi}{32})^{1/2}I(k)\frac{k^2 L_\gamma(\eta_{rec})V_A^2 \eta_{rec}}{3(1 + S_{rec})}$$

$$\approx 0.4\mu K \left(\frac{B_{-9}}{3}\right)^2 \left(\frac{l}{1000}\right)^2 I(\frac{l}{R_{rec}}). \tag{5.48}$$

where, $l = kR_{rec}$. The values of the cosmological parameters we have used are, $\Omega_\Lambda = 0.73$, $\Omega_m = 0.27$, $\Omega_b h^2 = 0.0224$ and $h = 0.71$. We also use the fit given in reference [24] to calculate $\eta_0 = 6000h^{-1}((1+a_{eq})^{1/2} - a_{eq}^{1/2})(1 -$

$0.0841 \ln(\Omega_m))/\Omega_m^{1/2}$, valid for flat universe. On scales where $kL_S > 1$ and $k\sigma > 1$, but $kL_\gamma(\eta_{rec}) < 1$, we get

$$\Delta T_P^{BB}(l) = T_0 \frac{\pi^{1/4}}{\sqrt{32}} I(k) \frac{5V_A^2}{3(k\sigma)^{1/2}}$$

$$\approx 1.2\mu K \left(\frac{B_{-9}}{3}\right)^2 \left(\frac{l}{2000}\right)^{-1/2} I(\frac{l}{R_{rec}}). \qquad (5.49)$$

The function $I^2(k)$ in the Eqs.(5.48) and (5.49) is a dimensionless mode-coupling integral given in our previous papers (cf. Eq. (7) of [14]). Analytic approximations to $I(k)$ exist for power-law spectra and for $k \ll k_c$ (as generally relevant even at high l; [14, 15]) For $n > -3/2$,

$$I^2(k) = \frac{28}{15} \frac{(n+3)^2}{(3+2n)} (\frac{k}{k_G})^3 (\frac{k_c}{k_G})^{3+2n}, \qquad (5.50)$$

and is dominated by the cut-off scale k_c. For $n < -3/2$ ([20]),

$$I^2(k) = \frac{8}{3}(n+3)(\frac{k}{k_G})^{6+2n} \qquad (5.51)$$

it is independent of k_c. Here we neglect a subdominant term of order $(k_c/k)^{3+2n} \ll 1$; note that $k \ll k_c$. If we consider a spectrum that is almost scale-invariant, say with $n = -2.9$, then we get $\Delta T_P^{BB}(l) \sim 0.16\mu K (l/1000)^{2.1}$ for scales larger than the Silk scale, and $\Delta T_P^{BB}(l) \sim 0.51\mu K (l/2000)^{-0.4}$, for scales smaller than L_S but larger than L_γ. Larger signals result for steeper spectra, $n > -2.9$ at the higher l end.

5.10 Non-Gaussianity from magnetic fields

Non-Gaussianity in the temperature anisotropy of the CMBR can arise due to a variety of reasons. Non-Gaussian signals from a cosmological magnetic field has focused on a homogeneous magnetic field [29]. In such cases, isotropy of space will be broken and could lead to an overall dipole anisotropy of the CMBR. In this section, such homogeneous magnetic fields will not be considered. On the other hand, we will focus on stochastic primordial fields which are statistically isotropic and homogeneous and investigate the possible non-Gaussian signals they could induce in the CMB.

Non-Gaussianity of the temperature anisotropy of the CMBR has recently drawn attention in the context of inflationary models. In such models, sub-Hubble scale linearized quantum fluctuations of the inflaton field lead to classical curvature perturbations on large scales. The Gaussian statistics of the initial quantum fluctuations lead to Gaussian statistics for the curvature perturbations. The resulting CMBR anisotropies will also have a Gaussian nature. If non-Gaussianity exists in these models, it can come only through higher order effects. The case of magnetic stresses is however a different. Even with a magnetic field with Gaussian probability distribution, non-Gaussianity in the CMBR can be induced. This is because, the magnetic stresses depend quadratically on the field. Hence, even for the field with Gaussian probability distribution, the related magnetic stresses have a non-Gaussian character even at the lowest order. Hence, the CMBR anisotropies induced by primordial fields will naturally have a non-Gaussian nature. Thus bounds on CMBR non-Gaussianity can put strong constraints on primordial magnetic fields, or lead to their detection.

Non-Gaussianity of temperature anisotropy could arise due to different reasons. The simplest case is when the 3-point function, also called the bi-spectrum, is non-zero. This provides an important and useful characterization of non-Gaussianity in the CMBR. It has more information than the single point probability distribution function (PDF). The simplest contribution to the CMBR bi-spectrum arises due to the magnetically induced Sachs-Wolfe type effects. We compute this contribution.

We have already discussed the nature of magnetic fields that we will consider. As mentioned in section 5.8 the magnetic field is assumed to be a Gaussian random field. On galactic scales and above, any velocity induced by Lorentz forces does not lead to appreciable distortion of the initial field [27]. So, the magnetic field simply redshifts away with expansion as $\mathbf{B}(\mathbf{x}, t) = \mathbf{b}_0(\mathbf{x})/a^2$. Here, $\mathbf{b}_0$ is the magnetic field at the present epoch.

The temperature anisotropies due to the magnetically induced Sachs-Wolfe effect is given in [30, 31]. These arise on large-angular scales, can be expressed as,

$$\frac{\Delta T}{T}(\boldsymbol{n}) = \mathcal{R}\,\Omega_B(\boldsymbol{x}_0 - \boldsymbol{n}D^*). \qquad (5.52)$$

Here $\Omega_B(\mathbf{x}) = \mathbf{B}^2(\mathbf{x}, t)/(8\pi\rho_\gamma(t)) = \mathbf{b}_0^2(\mathbf{x})/(8\pi\rho_0)$, where $\rho_\gamma(t)$ and ρ_0 are respectively the energy densities of the CMBR at times t and at the present epoch. (It is worth emphasizing at this point that the $\Delta T/T$ given above is on large-angular scales, just as in the case of the anisotropy power-spectrum.) An analytic estimate $\mathcal{R} = R_\gamma/20$ as the Sachs-Wolfe contribution is given in [30]. Here, $R_\gamma \sim 0.6$ is the fractional contribution of radiation energy density towards the total energy density of the relativistic components. The unit vector $\mathbf{n}$ gives the direction of observation and D^* is the (angular diameter) distance to the surface of last scatter. Earlier in this article, we had mentioned that the recombination era is a fairly well defined epoch. That is because there is a rapid transition from the plasma phase to the near neutral phase. Although this transition is not strictly instantaneous, for our purpose in this article, it is a fairly good approximation on large angular scales to assume it to be instantaneous. Additional contribution arising from integrated Sachs-Wolfe (ISW) to $\mathcal{R}$ is also possible [30].

The temperature fluctuations can be expanded in terms of the spherical harmonics to give $\Delta T(\boldsymbol{n})/T = \sum_{lm} a_{lm} Y_{lm}(\boldsymbol{n})$, where

$$a_{lm} = 4\pi \frac{1}{i^l} \int \frac{d^3k}{(2\pi)^3} \, \mathcal{R} \, \hat{\Omega}_B(\boldsymbol{k}) \, j_l(kD^*) Y_{lm}^*(\hat{\boldsymbol{k}}). \qquad (5.53)$$

Here $\hat{\Omega}_B(\boldsymbol{k})$ is the Fourier transform of $\Omega_B(\mathbf{x})$. Since $\Omega_B(\mathbf{x})$ is quadratic in $\mathbf{b}_0(\mathbf{x})$, we have $\hat{\Omega}_B(\boldsymbol{k}) = \frac{1}{(2\pi)^3} \int d^3s \, b_i(\boldsymbol{k}+\boldsymbol{s}) b_i^*(\boldsymbol{s})/(8\pi\rho_0)$, where now $b_i(\boldsymbol{k})$ is the Fourier transform of $\mathbf{b}_0(\mathbf{x})$.

We denote the bi-spectrum (3-point correlation function) by $B_{l_1 l_2 l_3}^{m_1 m_2 m_3}$. In terms of the a_{lm}'s it is given by,

$$B_{l_1 l_2 l_3}^{m_1 m_2 m_3} = <a_{l_1 m_1} a_{l_2 m_2} a_{l_3 m_3}>. \qquad (5.54)$$

From Eq. 5.53 we can express $B_{l_1 l_2 l_3}^{m_1 m_2 m_3}$ as

$$B_{l_1 l_2 l_3}^{m_1 m_2 m_3} = \mathcal{R}^3 \int \left[\prod_{i=1}^{3} (-i)^{l_i} \frac{d^3 k_i}{2\pi^2} j_{l_i}(k_i D^*) Y_{l_i m_i}^*(\hat{\boldsymbol{k}}_i) \right] \zeta_{123} \qquad (5.55)$$

with ζ_{123} defined as,

$$\zeta_{123} = <\hat{\Omega}_B(\boldsymbol{k}_1)\hat{\Omega}_B(\boldsymbol{k}_2)\hat{\Omega}_B(\boldsymbol{k}_2)>. \qquad (5.56)$$

The magnetic field itself is assumed to have a Gaussian probability distribution and to be non-helical in nature. Hence the power-spectrum, $M(k)$, specifies the statistical properties of the magnetic field completely. This spectrum is defined by the relation $< b_i(\mathbf{k}) b_j^*(\mathbf{q}) > = (2\pi)^3 \delta(\mathbf{k}-\mathbf{q}) P_{ij}(\mathbf{k}) M(k)$, where $P_{ij}(\mathbf{k}) = (\delta_{ij} - k_i k_j/k^2)$ is the projection operator ensuring $\nabla \cdot \mathbf{b}_0 = 0$. This gives $< \mathbf{b}_0^2 > = 2 \int (dk/k) \Delta_b^2(k)$, where $\Delta_b^2(k) = k^3 M(k)/(2\pi^2)$ is the power per logarithmic interval in k space residing in the stochastic magnetic field.

As in [21], we assume a power-law magnetic spectra, $M(k) = A k^n$ that has a cutoff at $k = k_c$, where k_c is the Alfvén-wave damping length-scale [27]. We fix A by demanding that the variance of the magnetic field smoothed over a 'galactic' scale, $k_G = 1 h \mathrm{Mpc}^{-1}$, (using a sharp k-space filter) is B_0. This gives, (for $n > -3$ and for $k < k_c$)

$$\Delta_b^2(k) = \frac{k^3 M(k)}{2\pi^2} = \frac{B_0^2}{2}(n+3) \left(\frac{k}{k_G} \right)^{3+n}. \qquad (5.57)$$

Since the energy density of the magnetic field is quadratic in the field strength, the 3-point correlation function of $\hat{\Omega}_B(\mathbf{k})$ involves a 6-point correlation function of the fields. The expression for ζ_{123} that appears in the integrand in equation (5.55) is given by,

$$\zeta_{123} = \delta(\mathbf{k}_1 + \mathbf{k}_2 + \mathbf{k}_3)\,\psi_{123}, \tag{5.58}$$

where

$$\psi_{123} = \frac{1}{(4\pi\rho_0)^3} \int d^3 s M(|\mathbf{k}_1 + \mathbf{s}|) M(s) M(|\mathbf{s} - \mathbf{k}_3|) F. \tag{5.59}$$

Here $F = \alpha^2 + \beta^2 + \gamma^2 - \alpha\beta\gamma$ with $\alpha = (\hat{\mathbf{s}} \cdot \widehat{\mathbf{s} + \mathbf{k}_1})$, $\beta = (\hat{\mathbf{s}} \cdot \widehat{\mathbf{s} - \mathbf{k}_3})$ and $\gamma = (\widehat{\mathbf{k}_1 + \mathbf{s}} \cdot \widehat{\mathbf{s} - \mathbf{k}_3})$, where the hat on a vector denotes its unit vector. This result has also been obtained in Ref. [32].

It is not possible to calculate the bi-spectrum analytically for the general case. However, for certain special cases, one can make an approximation and evaluate it. We calculate the bi-spectrum in two limits: (i) the 'equilateral' case for which the three l_i's are equal, and, (ii) the 'local isosceles' case for which $l_2 = l_3 \gg l_1$. In case(i), the presence of $j_{l_i}(k_i D^*)$ in Eq. (5.55), predominantly picks out those configurations in the Ω_B bi-spectrum for which $k_1 \sim k_2 \sim k_3$ In the local isosceles case (case ii), configurations in k values with $k_2 \sim k_3 \gg k_1$ are predominantly chosen.

The advantage with these specific configurations is that the mode coupling integral can be approximated in these cases following methods discussed in earlier works [14, 21]. In case (i), when $k_1 = k_2 = k_3$, we split the s-integral into the sub-ranges $0 < s < k_1$ and $s > k_1$. For each of these sub-ranges the mode-coupling integrand in Eq. 5.59 is approximated with $s \ll k_1$ and $s \gg k_1$, respectively. Similarly, in the second case, when $k_2 = k_3 \gg k_1$, the s-integral is now split into sub-ranges, $0 < s < k_1$, $k_1 < s < k_3$ and $s > k_3$. In each of these sub-ranges, we made the assumption that $s \ll k_1$, $k_1 \ll s \ll k_3$ and $s \gg k_3$, respectively. With this assumption, we approximate the mode coupling integrand in Eq. 5.59. For the possible values of n we also confine ourselves to spectral indices $-3 < n < -3/2$. In fact blue spectra are strongly constrained by a number of observations, particularly the gravitational wave limits of Ref. [33]. For numerical estimates we will focus on the case with $n \to -3$ which is a nearly scale invariant spectra. For case (i) we then get

$$\psi_{123} = \left(\frac{4}{3}\right)^4 \frac{\pi^7}{k_G^6} \frac{(n+3)^2(7-n)}{2(|n+1|)} \left(\frac{k_1}{k_G}\right)^{2n+3} \left(\frac{k_3}{k_G}\right)^n V_A^6, \tag{5.60}$$

while for case (ii) we have

$$\psi_{123} = \left(\frac{16}{3}\right)^3 \frac{\pi^7}{k_G^6} \frac{(n+3)^2}{|2n+3|} \left(\frac{k_1}{k_G}\right)^{2n+3} \left(\frac{k_3}{k_G}\right)^n V_A^6. \qquad (5.61)$$

Here we have defined V_A, the Alfvén velocity in the radiation era as

$$V_A = \frac{B_0}{(16\pi\rho_0/3)^{1/2}} \approx 3.8 \times 10^{-4} B_{-9}, \qquad (5.62)$$

with $B_{-9} \equiv (B_0/10^{-9}\text{Gauss})$.

We now express the delta function in equation (5.58) as $\delta(\boldsymbol{k}) = (1/(2\pi)^3) \int d^3x \exp(i\boldsymbol{k} \cdot \boldsymbol{x})$ and substitute for the exponential function in terms of spherical harmonics in Eq. (5.55). We then integrate over the angular parts of $(\boldsymbol{k}_1, \boldsymbol{k}_2, \boldsymbol{k}_3, \boldsymbol{x})$. The details of the algebra is along the lines of what is done for calculating the primordial bi-spectrum [34]. We can then write the bi-spectrum $B_{l_1 l_2 l_3}^{m_1 m_2 m_3}$, in terms of a reduced bi-spectrum $b_{l_1 l_2 l_3}$ as

$$B_{l_1 l_2 l_3}^{m_1 m_2 m_3} = \mathcal{G}_{m_1 m_2 m_3}^{l_1 l_2 l_3} \, b_{l_1 l_2 l_3} \qquad (5.63)$$

where

$$b_{l_1 l_2 l_3} = \left(\frac{\mathcal{R}}{\pi^2}\right)^3 \int x^2 dx$$

$$\times \prod_{i=1}^{3} \int k_i^2 dk_i \, j_{l_i}(k_i x) \, j_{l_i}(k_i D^*) \, \psi_{123} \qquad (5.64)$$

and we have introduced the Gaunt integral

$$\mathcal{G}_{m_1 m_2 m_3}^{l_1 l_2 l_3} = \int d\Omega \, Y_{l_1 m_1} Y_{l_2 m_2} Y_{l_3 m_3}. \qquad (5.65)$$

For case (i) (the equilateral case), we substitute Eq. (5.60) into Eq. (5.64) for the reduced bi-spectrum. We substitute Eq. (5.61) into Eq. (5.64) for the for case (ii). The integrals over k_2 can be done using $\int k_2^2 dk_2 j_{l_2}(k_2 x) j_{l_2}(k_2 D^*) = (\pi/2x^2)\delta(x - D^*)$, and the delta function picks out the value of x corresponding to the surface of last scatter. We are then left with integrals over k_1 and k_3 given by

$$b_{l_1 l_2 l_3} = \frac{\pi}{2} \left(\frac{\mathcal{R}}{\pi^2}\right)^3 V_A^6 \left[\int \frac{dk_3}{k_3} j_{l_3}^2(k_3 D^*) \left(\frac{k_3}{k_G}\right)^{n+3}\right]$$

$$\times \left[\int \frac{dk_1}{k_1} j_{l_1}^2(k_1 D^*) \left(\frac{k_1}{k_G}\right)^{2(n+3)}\right] C(n). \qquad (5.66)$$

For the equilateral case (case (i))

$$C(n) = \left(\frac{4}{3}\right)^4 \frac{\pi^7}{2} \frac{(n+3)^2(7-n)}{(|n+1|)}$$ (5.67)

where as for isosceles case (case (ii)) we have,

$$C(n) = \left(\frac{16}{3}\right)^3 \pi^7 \frac{(n+3)^2}{|2n+3|}.$$ (5.68)

The integral in Eq. (5.66) can be evaluated analytically in terms of Gamma functions. We may note that for power law spectra, the form of the integrals is the same as the usual Sachs-Wolfe term. The case of a nearly scale invariant case, $(n \approx -3)$ is of special interest. This is because such magnetic spectra are expected to arise in inflationary models for primordial magnetic field generation that we assume here [35]. For a purely Sachs-Wolfe contribution as in [30] gives $\mathcal{R} = R_\gamma/20 \sim 0.03$. For the equilateral case we then have

$$l_1(l_1+1)l_3(l_3+1)b_{l_1 l_2 l_3} \approx 2.3 \times 10^{-23} \left(\frac{n+3}{0.2}\right)^2 \left(\frac{B_{-9}}{3}\right)^6$$ (5.69)

while for the local-isosceles case we get

$$l_1(l_1+1)l_3(l_3+1)b_{l_1 l_2 l_3} \approx 1.5 \times 10^{-22} \left(\frac{n+3}{0.2}\right)^2 \left(\frac{B_{-9}}{3}\right)^6.$$ (5.70)

If the contribution of the Integrated Sachs-Wolfe term is included, the resulting signals may be expected to be larger. For $n \sim -3$, the numerical values of $l_1(l_1+1)l_3(l_3+1)b_{l_1 l_2 l_3}$ is higher for the local isosceles case as compared to the equilateral case by a factor ~ 6.4 The origin of this ratio can be traced to the ratio of ψ_{123} for these two cases. From equations 5.60 and 5.61 we see that $\psi_{123(local)} = \psi_{123(equil)}(96 \mid n+1 \mid)/(\mid 2n+3 \mid (7-n))$ which for $n \sim -3$ comes out to be $\sim 6.4\psi_{123(equil)}$.

It is instructive to compare these values for the reduced bi-spectrum, with those at large angular scales, which arises due to nonlinear terms in the gravitational potential, characterized by f_{NL} (cf. [36]). The value of $l_1(l_1+1)l_3(l_3+1)b_{l_1 l_2 l_3}$ for the latter is about $4 \times 10^{-18} f_{NL}$. Thus the magnetically induced non-Gaussian signal, due to the purely Sachs-Wolfe effect (with $\mathcal{R}$ as in [30]), is a factor of about a few times 10^4 smaller than the standard signal predicted in inflationary models with $f_{NL} \sim 1$. If observations constrain $l_1(l_1+1)l_3(l_3+1)b_{l_1 l_2 l_3} < 4 \times 10^{-16}$, for $f_{NL} < 100$ (for example say from WMAP experiment [37]), then we have a limit of $B_0 < 35$ nano Gauss on the strength of any primordial field with a nearly scale invariant spectrum. Including the contribution from ISW would lead to stronger limits.

For the standard primordial contribution to the bi-spectrum on large angular scales (for scale invariant potential perturbations) the l_i dependence is the same as that we get here, in both cases (i) and (ii), for magnetically induced Sachs-Wolfe effect (for scale-invariant magnetic spectra) (cf. [34, 36]). In all these cases $l_1(l_1 + 1)l_3(l_3 + 1)b_{l_1 l_2 l_3}$ is independent of l's (see Eq. (17) and (18) and Ref. [36]). Thus the bounds on f_{NL} got from the WMAP data for the standard 'local' non-Gaussianity are indeed useful to set constraints on B_0. The WMAP limits use the much larger range of l values than the range for which the Sachs-Wolfe contribution is important. However, the limit on B_0 depends only very weakly ($B_0 \propto f_{NL}^{1/6}$) on the exact observational limit on f_{NL}. Thus limits on B_0 arrived at here are expected to be reasonably robust.

Using bi-spectrum as a probe for stochastic primordial magnetic fields, has a unique advantage over the power spectrum. The magnetically induced signal is fundamentally non Gaussian. Thus it can more easily be distinguishable in the bi-spectrum. This is because the bi-spectrum arising due to magnetic contribution can in principle dominate that arising from models of inflation with a small enough f_{NL}. The problem with the power spectrum arising due to magnetic contribution is that it is generally subdominant to those arising from inflation generated curvature perturbations, for nano Gauss fields and scale invariant spectra.

References

[1] Penzias, A. A. and R.W. Wilson, *Astrophys. J.* **142**, 419, (1965)

[2] See, *http://astrophysics.arc.nasa.gov/ mway/AMES-CMB.pdf*

[3] Dodelson, S., *Modern Cosmology*, Elsevier, 2003

[4] Challinor, A. and Hiranya Peiris, *Lecture Notes on the Physics of Cosmic Microwave Background Anisotropies.* (astro-ph/0903.5158)

[5] Samtleben, D., S. Staggs and B. Winstein, *The Cosmic Microwave Background Radiation for Pedestrians:* (astro-ph/0803.0834)

[6] Padmanabhan, T., *Structure formation in the universe*, Cambridge University Press

[7] Padmanabhan, T., *Theoretical Astrophysics, Galaxies and Cosmology*, Cambridge University Press, Cambridge 2002, vol. III

[8] Durrer, R., *The Cosmic Microwave Background*, Cambridge University

Press, Cambridge 2008

[9] Giovannini, G., *A Primer on the Physics of the Cosmic Microwave Background*, World Scientific, 2008

[10] Subramanian, K., *Current Science*, 2005, **88**, 1068

[11] Kosowsky, A., *New Astron.Rev.* 1999, **43** 157

[12] Hu, W. and White, M., *New Astronomy* 1997 **2** 323

[13] Rees, M. J., *Astrophys. J.* 1968 **153** L1

[14] T. R. Seshadri and K. Subramanian, Phys. Rev. Lett. **87**, 101301 (2001)

[15] A. Mack, T. Kashniashvili and A. Kosowski, Phys. Rev. **D 65**, 123004 (2002)

[16] A. Lewis, Phys. Rev. **D 70**, 043011 (2004)

[17] Sunyaev, R. A. and Zeldovich, Y. B., 1972, Comm. Astrophys. Space Phys., 4, 173

[18] A. A. Ruzmaikin, A. Shukurov and D. Sokoloff, *Magnetic Fields of Galaxies*, Kluwer, Dordrecht (1988); L. Mestel, *Stellar Magnetism*, Oxford UP, Oxford, (1998); R. Beck, A. Brandenburg, D. Moss, A. Shukurov and D. Sokoloff, Ann. Rev. A. & A., **34**, 155 (1996).

[19] A. Brandenburg and K. Subramanian, Submitted to Physics Reports (astro-ph/0405052,).

[20] K. Subramanian and J. D. Barrow, Mon. Not. R. Astron. Soc., **335**, L57 (2002)

[21] K. Subramanian, T. R. Seshadri and J. D. Barrow, Mon. Not. R. Astron. Soc. **344**, L31 (2003)

[22] R. Durrer, P. G. Ferreira and T. Kashniashvili, Phys. Rev. **D 61**, 043001 (2000)

[23] K. Subramanian and J. D. Barrow, Phys. Rev. Lett. **81**, 3575 (1998)

[24] W. Hu and M. White, Phys.Rev. **D 56**, 596 (1997)

[25] D. N. Spergal et. al. astro-ph/0302209

[26] W. Hu and N Sugiyama, Apj **444**, 489 (1995)

[27] K. Jedamzik, V. Katalinic, and A. Olinto, Phys. Rev. **D57**, 3264 (1998).

[28] K. Subramanian and J. D. Barrow, Phys.Rev. **D58** 083502 (1998).

[29] G. Chen, P. Mukherjee, T. Kahniashvili, B. Ratra and Yun Wang, Astrophys.J., **611**, 655 (2004); P. D. Naselsky, L-Y. Chiang, P. Olesen and O. V. Verkhodanov, Astrophys. J., **615**, 45 (2004); A. Bernui and W.S. Hipolito-Ricaldi, Mon. Not. R. Astron. Soc., **389**, 1453 (2008); T. Kahniashvili, G. Lavrelashvili and B. Ratra, Phys. Rev. D., **78**, 063012 (2008)

[30] M. Giovannini, PMC Physics A, **1**:5 (2007); (doi:10.1186/1754-0410-1-5)

[31] D. Paoletti, F. Finelli and F. Paci, arxiv:0811.0230 (2008)

[32] I. Brown, and R. Crittenden, Phys. Rev. D, **72**, 063002 (2005)

[33] C. Caprini, R. Durrer, Phys. Rev. Đ, **65**, 3517 (2002)

[34] J. R. Fergusson and E. P. S. Shellard, Phys. Rev. D, **76**, 083523 (2007)

[35] M. Turner, L. M. Widrow, Phys. Rev. D. **37**, 2743–2754 (1988); B. Ratra, ApJ, **391**, L1 (1992); L. M. Widrow, Rev. Mod. Phys., **74**, 775–823 (2002); M. Giovannini, "String theory and fundamental interactions", eds. M. Gasperini and J. Maharana, Lecture Notes in Physics, Springer, Berlin/Heidelberg (2007) (arXiv:astro-ph/0612378)

[36] A. Riotto, *The quest for Non-Gaussianity*, Lect. Notes Phys., **738**, 305 (2008)

[37] E. Komatsu et al., Astrophys. J. Suppl. **180**, 330 (2009)

Chapter 6

Quantum corrections to Bekenstein-Hawking entropy

S. Shankaranarayanan

School of Physics,
Indian Institute of Science Education and Research-Trivandrum,
CET campus, Thiruvananthapuram 695 016, India
E-mail: *shanki@iisertvm.ac.in*

Abstract: There are strong indications that black holes are thermodynamic objects having, in particular, well-defined notions of temperature and entropy. The entropy (temperature) is finite (non-zero) only if the matter fields propagating in the black hole background are treated as quantum. In the search for a quantum theory of gravity, black hole thermodynamics has been used as a theoretical laboratory. Some of the fascinating issues arising in the struggle to understand black hole thermodynamics are (i) statistical origin of black hole entropy, (ii) Holographic paradigm and (iii) information paradox. These are related to what seems to be a deep and fundamental problem – the counting of relevant degrees of freedom. In this article, we ask two questions: (i) Using simple techniques, can we know whether an approach to black hole entropy contains semi-classical or quantum degrees of freedom? (ii) Is it possible to identify generic quantum corrections to Bekenstein-Hawking entropy? Using naive-dimensional analysis, we show that it is possible to differentiate between different approaches and argue that the quantum corrections should have different scaling behaviour. We then show that the approach of entanglement of modes across the horizon leads to generic quantum corrections and that the microscopic degrees of freedom that lead to Bekenstein-Hawking entropy and sub-leading corrections are different.

6.1 Paddy

When Sriramkumar and Seshadri asked me to write an article dedicated to *Paddy*, my initial thought was to write on Trans-Planckian inflation since, he was the first person to look at the plausible quantum gravitational effects in CMB [1]. However, on second thought, I decided to write on quantum corrections to black hole entropy. This was partly due to the fact that, in the last decade, Paddy has spent sufficiently long time on unearthing links between thermodynamics and gravity [2, 3, 4, 5, 6, 7, 8]. But it is also because, while I consider understanding quantum corrections to black hole entropy to be the key to black hole entropy and thermodynamics — at least to my understanding — Paddy seem to think the opposite. So, I hope that Paddy will find it interesting and that this article will act as a catalyst to initiate a serious discussion on issues and the importance of quantum corrections to black hole entropy.

6.2 Prologue

Any review on black hole thermodynamics starts with an extensive discussion on Bekenstein-Hawking entropy and Hawking radiation with a little discussion on entropy and the subtleties in quantifying it. Hence, in this review, we have decided to discuss more about entropy following the classic review of Wehrl [9]. There are two reasons for this: (i) The connection between black hole mechanics and ordinary thermodynamics was established by Hawking by showing that the surface gravity of the event horizon equals 2π times the ordinary temperature of the radiation emitted by the black hole to infinity [10]. The original derivation of the laws of black hole thermodynamics has nothing in common with statistical mechanics, but there is a general belief that a connection nevertheless exists at the quantum level. At the leading order, several approaches using completely different microscopic degrees of freedom lead to Bekenstein-Hawking entropy. Hence, it is important to understand why all these diverse approaches lead to identical entropy. (ii) Using thermodynamical reasoning, one expects entropy to be a state function of the system. This state connotation is a source of ambiguities, since in-equivalent notions of the system state are used in the description of physical systems [9] and it plays a significant role when we discuss sub-leading corrections to the black hole entropy.

In the following section (6.3), we will discuss entropy and its notions.

We also discuss von Neumann's thought experiment and the assumptions involved in obtaining the operator uniquely and put-forward in the context of black hole entropy. In Sec. (6.4), we discuss an intriguing feature of black hole entropy — semi-classical entropy of black hole in any gravity theory closely follow the form of classical gravity action. Based on this observation, we argue that the quantum corrections should have different scaling behaviour. Sec. (6.5) discusses the importance of quantifying quantum corrections. In Sec. (6.6), we provide a heuristic picture of the link between the entanglement entropy and black hole entropy. We then provide a rapid review of entanglement entropy and the quantifying tool for the same. In Sec. (6.7) we discuss the procedure and assumptions to compute the entanglement entropy of a scalar field in black hole space-times. In Secs. (6.8, 6.9), we provide the key results in microcanonical and canonical ensemble approaches, respectively. Finally, we conclude in Sec. (6.10), summarising our results and speculating on future directions.

6.3 Entropy and the choice of system states

In physics, entropy is regarded as a measure of the degree of randomness and the tendency of physical systems to become less organised leading to decrease in the information about the physical system [9]. Hence, the notions of entropy, information and uncertainty are intertwined and hard to differentiate. While entropy and uncertainty are equivalent; measures of ignorance — the complementary notion of information — quantifies the ability of observers to make reliable predictions about the system. In other words, the more one is aware about chances of a concrete outcome, the lower is the uncertainty of this outcome. Normally, the growth of uncertainty is identified with an increase of the entropy which in turn is interpreted as the loss of information.

Entropy is a derived quantity and it does not show up in any fundamental equation of motion. However, in any physical theory, entropy takes a unique position among other physical quantities. This is due to the fact that the entropy relates the macroscopic and microscopic structure of a system and determines its behaviour in (near) equilibrium [9]. Technically, it provides non-trivial information about the microscopic structure of the system through the Boltzmann relation

$$S = k_B \ln \Omega \equiv -k_B \ln \mathcal{P} \tag{6.1}$$

where k_B is the Boltzmann constant and Ω is the total number of possible equiprobable) microstates that imply the prescribed macroscopic (e.g. thermodynamical) behaviour corresponding to a fixed value of S.

The Boltzmann relation links entropy of the (thermodynamical) system with the probability $\mathcal{P} = 1/\Omega$ that an appropriate "statistical microstate" can occur. Here, $\ln \mathcal{P}$ may be interpreted as the degree of uncertainty in the trial experiment. Another interpretation of $\ln \mathcal{P}$ is that of a measure of information produced when one microstate is chosen from the set, all choices being equally likely.

However, due to in-equivalent choice of the system state, the interpretation of entropy as a measure of uncertainty acquires ambiguities. For instance, a phase-space point determines the state of a classical dynamical system or the macroscopic notion of a thermodynamical state in its classical and quantum versions [9]. In order to have unambiguity, one must have a catalogue of theoretical predictions and their corresponding observables.

In the case of quantum theory, which is our interest in this review, the representation of quantum states in terms of wave functions or density operators does provide experimentally verifiable information about the system. Observables correspond to self-adjoint operators and statistical operators are associated with the states. Adopting the state notion to the Hilbert space language of quantum theory, it is clear that normalised wave functions and density operators allow to extend the notion of entropy to certain functionals of the state of the quantum system and hence, the interpretation of entropy as a measure of uncertainty in a quantum state can acquire an unambiguous meaning.

von Neumann associated an entropy quantity to a statistical operator [11]. While the operator is well-known, it is less-known what assumptions are involved and how the operator is obtained. In the following subsection we will discuss, in detail, the original thought experiment used by von Neumann to obtain the operator by highlighting the assumptions [11]. We will then discuss the limitations in the use of von Neumann operator in some approaches to black hole entropy.

6.3.1 *Thought experiment by von Neumann*

Consider a gas of $N(\gg 1)$ molecules in a rectangular box Y. Suppose that the gas behaves like a quantum system and is described by a statistical operator D which is a mixture $\lambda|\phi_1\rangle\langle\phi_1| + (1 - \lambda)|\phi_2\rangle\langle\phi_2|$, $|\phi_i\rangle \equiv \phi_i$ is a state vector ($i = 1, 2$). Let us assume that λN molecules are in the pure state ϕ_1 and $(1 - \lambda)N$ molecules are in the pure state ϕ_2. On the basis of thermodynamics, if ϕ_1 and ϕ_2 are orthogonal, then there is a wall which is completely permeable for the ϕ_1-molecules and isolating for the ϕ_2-molecules.

Let us now add an equally large empty rectangular box Y' to the left of the box Y and we replace the common wall with two new walls. Wall (a), the one to the left is impenetrable, whereas the one to the right, wall (b), lets ϕ_1-molecules through but keeps back ϕ_2-molecules. Now, add a third wall (c) opposite to (b) which is semi-permeable, transparent for the ϕ_2-molecules and impenetrable for the ϕ_1-ones. We now push slowly (a) and (c) to the left, maintaining their distance. During this process ϕ_1-molecules are pressed through (b) into Y' and ϕ_2-molecules diffuse through wall (c) and remain in Y. No work is done against the gas pressure, hence, no heat is developed. If we replace the walls (b) and (c) with a rigid absolutely impenetrable wall and remove (a), we restore the boxes Y and Y' and succeed in the separation of ϕ_1-molecules from the ϕ_2-ones without any work being done, without any temperature change and without evolution of heat.

The entropy of the original D-gas (with density N/V) must be the sum of the entropies of the ϕ_1- and ϕ_2-molecules (with densities $\lambda N/V$ and $(1-\lambda)N/V$, respectively). If we compress the gases in Y and Y' to volumes λV and $(1-\lambda)V$, respectively, keeping the temperature T constant by means of a heat reservoir, the entropy change amounts to $k_B \lambda N \ln \lambda$ and $k_B (1-\lambda) N \ln(1-\lambda)$, respectively. Finally, mixing ϕ_1- and ϕ_2-gases of identical density we obtain a D-gas of N molecules in a volume V at the original temperature. So, the entropy $S_0(\psi, N)$ of a ψ-gas of N molecules (in a volume V and at the given temperature), is given by

$$S_0(\phi_1, \lambda N) + S_0(\phi_2, (1-\lambda)N) = S_0(D, N) + k_B \lambda N \ln \lambda \qquad (6.2)$$
$$+ k_B (1-\lambda) N \ln(1-\lambda)$$

Assuming that $S_0(\psi, N)$ is proportional to N and dividing by N, we get,

$$\lambda S(\phi_1) + (1-\lambda) S(\phi_2) = S_0(D, N) + k_B \lambda \ln \lambda + k_B (1-\lambda) \ln(1-\lambda) \quad (6.3)$$

Using the fact that ϕ_1 and ϕ_2 are orthogonal, we get,

$$S\left(\sum_i \lambda_i |\phi_i\rangle\langle\phi_i|\right) = \sum_i \lambda_i S\left(|\phi_i\rangle\langle\phi_i|\right) - k_B \sum_i \lambda_i \ln \lambda_i \qquad (6.4)$$

The above equation reduces the determination of the (thermodynamical) entropy of a mixed state to that of pure states. However, the decomposition $\sum_i \lambda_i |\phi_i\rangle\langle\phi_i|$ of a statistical operator is not unique even if $\langle\phi_i, \phi_j\rangle = 0$ for all i and j. On the other hand, von Neumann wanted to avoid degeneracy of the spectrum of a statistical operator. To obtain a unique spectral operator, von Neumann assumed that any measurement process should satisfy second-law of thermodynamics by which he showed that $S(|\phi\rangle\langle\phi|)$ is independent of the state vector ϕ so that

$$S\left(\sum_i \lambda_i |\phi_i\rangle\langle\phi_i|\right) = -k_B \sum_i \lambda_i \ln \lambda_i = -k_B Tr\left[\hat{\rho}\ln(\hat{\rho})\right] \qquad (6.5)$$

Let us now recapitulate von Neumann's argument of the thermodynamical entropy of a statistical operator D. First of all, he assumed that $S(D)$ is a continuous function of D. He carried out a reversible process to obtain the mixing property (6.3) for orthogonal pure states, and he concluded (6.4). He referred to the second law and showed that $S(|\phi\rangle\langle\phi|)$ is independent of the state vector $|\phi\rangle$ which leads to (6.5).

But what has this digression to do with black hole entropy? Firstly, in any approach to black hole entropy, it is taken for granted that the assumption — $S(D)$ is a continuous function of D — is valid. It is unclear whether such an assumption can hold especially in the approaches inspired by Wheeler's *it from bit* [12].

To elaborate this, let us consider a two dimensional finite *floating lattice* with plaquettes approximately the size of a Planck area (l_{Pl}^2) covering the spherical horizon of a macroscopic[1] non-rotating four dimensional black hole. Assume that (i) binary variables are distributed randomly on this lattice and (ii) the size of the lattice is characterised by a finite large even integer p.

Usually what is done is to assume the dimensionality of the Hilbert space of quantum states to be $N(p) = 2p$, and hence, the number of degrees of freedom characterising the horizon is taken to be $N \equiv \log N(p) = p\log 2$. Assuming that $p \gg 1$; in the limit of very large p, the lattice can be taken to approximate the macroscopic horizon of the black hole. One would then expect that the classical horizon area would satisfy $\mathcal{A}_{\mathrm{H}}/l_{\mathrm{Pl}}^2 = \xi p$ where $\xi = \mathcal{O}(1)$. For the choice $\xi = 4\log 2$, one obtains for the entropy $\mathcal{S}_{\mathrm{BH}} \equiv N = \mathcal{A}_{\mathrm{H}}/(4l_{\mathrm{Pl}}^2)$. While it is natural to assume the dimensionality of the Hilbert space of quantum states to be $N(p) = 2p$, it is not clear whether the assumption for the von Neumann entropy operator $S(D)$ to be a continuous function of D is valid.

Secondly, the uniqueness of von Neumann operator was established by carrying out reversible process. It is unclear whether the use of this opera-

---

[1]classical area of the horizon $\mathcal{A}_{\mathrm{H}} \gg l_{\mathrm{Pl}}^2$

tor is valid in extremal or near-extremal black-holes [13, 14]. The Hawking temperature for these black-holes is zero or tiny and any process which leads to these black-holes is irreversible [15, 16]. Using the fact that the near-horizon geometry of 4-dimensional extremal black-holes has the form $AdS_2 \times S_2$, there has been an interest in using von Neumann operator to understand the AdS/CFT conjecture [14]. However, as mentioned above, the choice of von Neumann entropy in these cases need a thorough investigation before looking at the broader implications.

Having discussed in detail the assumptions leading to von Neumann and its implications for some approaches to black hole entropy, in the next section, we ask the following question: Using simple techniques, can we know whether an approach to black hole entropy contains semi-classical or quantum degrees of freedom? Using naive-dimensional analysis, we show that it is possible to differentiate between different approaches and argue that the quantum corrections should have different scaling behaviour.

6.4   An intriguing feature of black hole entropy

The area (as opposed to volume) proportionality of black hole entropy has engaged all the researchers in this field, however, there is an intriguing feature of black hole entropy which has received little discussion in the literature.

Let us consider Einstein-Hilbert action in 4-dimensions:

$$S_{\text{EH}} = \frac{M_{\text{Pl}}^2}{2} \int d^4x \sqrt{-g}\, R \tag{6.6}$$

where $M_{\text{Pl}}^2 \equiv 1/(8\pi G)$ is the reduced Planck mass. Naive dimensional analysis [17] of the above action leads to

$$S_{\text{EH}} \propto M_{\text{Pl}}^2 \times [L]^2 \tag{6.7}$$

The Bekenstein-Hawking entropy for 4-dimensional black-holes [18, 19] is given by

$$\mathcal{S}_{\text{BH}} = \frac{k_B}{4} \frac{\mathcal{A}_{\text{H}}}{l_{\text{Pl}}^2} \equiv \frac{k_B}{4} M_{\text{Pl}}^2 \mathcal{A}_{\text{H}} \tag{6.8}$$

Naive dimensional analysis of the above entropy leads to

$$\mathcal{S}_{\text{BH}} \propto M_{\text{Pl}}^2 \times [L]^2 \tag{6.9}$$

An attentive reader might now realise that the Einstein-Hilbert action (6.7) and Bekenstein-Hawking entropy (6.9) have the same dimensional dependence, indicating that the entropy of black-holes seem to directly follow the gravity action. However, a sceptic would say that this relation is only true for Einstenian gravity and may not work for a general gravity action.

Let us now consider an arbitrary gravity action, containing higher-order derivatives, in D-dimensional space-time:

$$S_{\text{Gen}} = \frac{1}{16\pi G_D} \int d^D x \sqrt{-g} \left[R + \alpha\, F(R^2) + \beta\, G(R^3) + \cdots \right]$$

$$= \frac{M_{\text{P}}^{(D-2)}}{16\pi} \int d^D x \sqrt{-g} \left[R + \alpha\, F(R^2) + \beta\, G(R^3) + \cdots \right] \quad (6.10)$$

where G_D is the D-dimensional Newton's constant, M_{P} is the D-dimensional effective Planck mass given by $G_D \sim M_{\text{P}}^{-(D-2)}$, $F(R^2)$ is an arbitrary function containing $R^2, R_{AB}R^{AB}, R_{ABCD}R^{ABCD}$ terms, $G(R^3)$ contains all the cubic terms of R, and α, β are dimensionful constants[2]. Let us again do a naive dimensional analysis for the above action:

$$S_{\text{Gen}} \propto M_{\text{P}}^{(D-2)} \times [L]^{D-2} \left[1 + \frac{\alpha}{[L]^2} + \frac{\beta}{[L]^4} + \cdots \right] \quad (6.11)$$

The semi-classical Noether charge entropy for the above generalised action is given by [23] (see also, [24, 25, 26, 27, 28, 29, 21, 22]):

$$S_{\text{NC}} = M_{\text{P}}^{(D-2)} \frac{\mathcal{A}_{\text{D}}}{4} \left[1 + \alpha \mathcal{A}_{\text{D}}^{2/(D-2)} + \beta \mathcal{A}_{\text{D}}^{2/(D-2)} + \cdots \right]. \quad (6.12)$$

Naive dimensional analysis of the above entropy leads to:

$$S_{\text{NC}} \propto M_{\text{P}}^{(D-2)} \times [L]^{D-2} \left[1 + \frac{\alpha}{[L]^2} + \frac{\beta}{[L]^4} + \cdots \right] \quad (6.13)$$

This observation indicates that the (semi-classical) black hole entropy in any gravity theory follow the form of the classical action of gravity. So, what are the physical consequences of this observation? Firstly, it clearly points that if any approach to black hole entropy predicts the same dimensional form as that of the classical action of gravity, then this approach *only* provides the semi-classical part of the theory and does not provide information about the quantum structure of gravity. This might seem a strong assertion, however, it would be an even stronger claim if one says that the quantum corrections to the semi-classical black hole entropy follow the same dimensionality of the classical action. [In the next section, we elaborate this point by taking a simple example of ideal gas.] This also implies that the recent results from string theory using the attractor mechanism for extremal black-holes also provide information about the semi-classical

---

[2]Note that the above action includes Lovelock gravity whose equations of motion, like Einstenian gravity, are quasi-linear [20, 21, 22].

aspect of gravity and does not necessarily provide information on the quantum aspect of gravity (for a recent review, see Ref. [30]).

Secondly, this suggests that, if we obtain black hole entropy from the complete quantum theory of gravity, the entropy will include terms which will not follow the form of the gravity action. At least two of the approaches to black hole entropy do seem to agree with this conjecture. For instance, using the leading order correction to Cardy formula, Carlip [31] showed the conformal field theory calculation predicts logarithmic corrections to Bekenstein-Hawking entropy. Similarly, in the quantum geometry approach, Kaul and Majumdar [32] showed that Hilbert space of the horizon of spherically symmetric space-time is $2d\,SU(2)_k$ Wess-Zumino model and obtained the generic log corrections. In the next section, we will discuss the importance of quantifying the corrections to Bekenstein-Hawking entropy by comparing with that of ideal gas.

6.5 Quantum corrections

To understand the importance of quantum corrections, let us compare black-holes with an ideal gas. Black-holes are the simplest objects; they can be described completely by few parameters like mass and charge. Ideal gas as the name suggests, is an idealisation that the gas molecules do not have any interaction, however, to describe them, we need to have information about the position and momenta of N particles.

The classical entropy of mono-atomic ideal gas (i. e. by assuming that the gas molecules are distinguishable) is given by

$$\frac{S_{\text{ideal}}}{k_B\,N} = \ln\left(V\,T^{3/2}\right) + C_0 \qquad (6.14)$$

where V is the volume, T is the temperature and C_0 is an arbitrary constant which cannot be obtained using classical equations of motion.

Supposing we assume that all atoms move independently, it is possible to calculate the number of quantum states available in an ideal gas and using this, one can obtain the exact expression for the entropy of a mono-atomic ideal gas:

$$\frac{S_{\text{ST}}}{k_B\,N} = \ln\left(V\,T^{3/2}\right) + \frac{1}{2}\ln\frac{M^3}{N^5} + \frac{3}{2}\ln\left(\frac{4\pi k_B}{3\hbar^2}\right) \qquad (6.15)$$

This is known as Sackur-Tetrode equation [33]. The significance of Sackur-Tetrode equation is that the first term matches with classical expression

(6.14) while the last two terms — the second term which is constant for a particular gas and the third term which is a universal constant — could not have been foreseen from classical thermodynamics. It is also interesting to note that the last term diverges as $\hbar \to 0$.

Why is Sackur-Tetrode equation relevant for black hole entropy? Firstly, like Sackur-Tetrode equation for ideal gas, in a pure classical description (of gravity and matter), black-holes have infinite entropy[3]. The finiteness of the black hole entropy requires that the matter and(or) gravity have quantum description.

Secondly, in the case of classical and quantum computation of ideal gas entropy, we know precisely what are the degrees of freedom which contribute to the ideal gas entropy. In the classical case, we assumed the gas to be distinguishable mono-atomic gas while in the quantum case, we calculated the number of quantum states available in an ideal gas, supposing all atoms to be moving independently from each other in a well defined volume. For example, we can know the precise number of quantum states for 1 mole of argon, at a temperature of 300 K and a pressure of 1 bar. It turns to be $\sim 10^{4,870000,000000,000000,000000}$ [34].

However, it is still unclear how S_{BH} concords with the standard view of the statistical origin? More importantly, what are *the* microscopic degrees of freedom leading to black hole entropy? Currently, there are several approaches starting from counting states (by assuming fundamental structures) [35, 36, 37] to Noether charge [38, 39, 40, 23, 41, 42, 43, 44, 45, 46, 47, 48, 49].

Thirdly, the second and third terms in the RHS of (6.15) are quantum corrections to the classical entropy which do not have any dependence on the macroscopic quantities like temperature, pressure or volume. It is needless to say that the dependence of these terms could not have been foreseen from classical thermodynamics or physical arguments. In the same manner, it would be impossible to predict the quantum corrections to the Bekenstein-Hawking entropy. If one uses symmetry arguments based on classical gravity action, then what we might obtain will be proportional to the form obtained using the naive dimensional analysis (6.8, 6.12).

Lastly, in the case of ideal gas, the degrees of freedom are uniquely identified and lead to the classical thermodynamic entropy. However, in the case of black-holes, as mentioned above, we seem to have several approaches with completely different degrees of freedom leading to identical

---

[3]In the $\hbar \to 0$, the constant last term in the RHS of (6.15) is infinite.

semi-classical Bekenstein-Hawking entropy (6.8). Although, none of these approaches can be considered to be complete; all of them — within their domains of applicability — by counting certain microscopic states yield (6.8).

The above discussion raises an important question: *Is it sufficient for an approach to reproduce (6.8) or does it need to go beyond the Bekenstein-Hawking entropy?* As we know, S_{BH} is a semi-classical result and there are strong indications that Eq. (6.8) is valid for large black holes [i.e. $\mathcal{A}_H \gg \ell_{Pl}$]. However, it is not clear whether this relation will continue to hold for the Planck-size black-holes. Besides, there is no reason to expect S_{BH} to be the whole answer for a correct theory of quantum gravity. In order to have a better understanding of black hole entropy, it is imperative for any approach to go beyond S_{BH} and identify the sub-leading corrections.

This raises a related question: *Are the quantum degrees of freedom (DOF) that contribute to S_{BH} and its sub-leading corrections, identical or different?* In general, the quantum DOF can be different. However, several approaches in the literature [31, 50, 26, 30] that do lead to sub-leading corrections either assume that the DOF are identical or do not *disentangle* DOF contribution to S_{BH} and the sub-leading corrections.

In the rest of this article, we show that the approach of entanglement predicts generic quantum corrections to Bekenstein-Hawking entropy. We also show that the quantum DOF that contribute to the Bekenstein-Hawking entropy and its sub-leading corrections are different. We also show that it is possible to *disentangle DOF contributions.* In the next section, we discuss the basics of quantum entanglement and why quantum entanglement might be relevant for the black hole entropy.

6.6 Quantum entanglement

6.6.1 *Relevance of entanglement for black hole entropy*

Black-hole entropy, or broadly thermodynamics, is expected to provide information about the structure of quantum gravity. In the absence of a workable theory of quantum gravity it is then necessary to have an approach which will incorporate all features of quantum theory of gravity, yet does not depend on the details of any approach to quantum gravity. Quantum entanglement is one such approach.

Among other reasons, we list below three primary reasons for the relevance of entanglement for black hole entropy. Firstly, entanglement like black hole entropy is a quantum effect with no classical analogue. Secondly, entanglement entropy and black hole entropy are associated with the existence of horizon. To elaborate this, let us consider a scalar field on a background of a collapsing star. Before the collapse, an outside observer, at least theoretically, has all the information about the collapsing star. Hence, the entanglement entropy is zero. During the collapse and once the horizon forms, S_{BH} is non-zero. The outside observers at spatial infinity do not have the information about the quantum degrees of freedom inside the horizon. Thus, the entanglement entropy is non-zero. In other words, both the entropies are associated with the existence of horizon[4]. Thirdly, entanglement is a generic feature of a quantum theory and hence should be present in any quantum theory of gravity. Although, the results presented below do not involve quantisation of gravity, they do have implications for the full theory.

With this background, we now review the basics of quantum entanglement and then go on to the basic setup of calculating entanglement entropy for black hole space-times in Einstenian gravity.

6.6.2 *Rapid review of entanglement*

Consider a quantum mechanical system, which can be decomposed into two subsystems u and v, as shown below [5]:

$$\boxed{\text{System}} = \boxed{\text{Subsystem } u} + \boxed{\text{Subsystem } v}$$

Correspondingly, the Hilbert space of the system is a Kronecker product of the subsystem Hilbert spaces:

$$\mathcal{H} = \mathcal{H}_u \otimes \mathcal{H}_v . \tag{6.16}$$

If $|u_i\rangle$ and $|v_j\rangle$ are eigen-bases in $\mathcal{H}_u$ and $\mathcal{H}_v$ respectively, then $|u_i\rangle \otimes |v_j\rangle$ form an eigen-basis in $\mathcal{H}$, in terms of which a generic wave-function $|\Psi\rangle$ in $\mathcal{H}$ can be expanded:

$$|\Psi\rangle = \sum_{ij} d_{ij}|u_i\rangle \otimes |v_j\rangle \in \mathcal{H} . \tag{6.17}$$

---

[4]It is important to understand that it is possible to obtain a non-vanishing entanglement entropy for a scalar field in flat space-time by artificially creating a horizon [40]. However, in the case of black hole, the event horizon is a physical boundary beyond which the observers do not have access to information.

[5]for details, we refer the reader to the review [51].

Note however, that in general, $|\Psi\rangle$ cannot be factorised into two wave-functions, one in u and the other in v:

$$|\Psi\rangle \neq |\Psi_u\rangle \otimes |\Psi_v\rangle . \qquad (6.18)$$

Such states are called *Entangled States* or *EPR states*. The ones that can be factorised, on the other hand, are called *Unentangled States*. For example, for a system with two spins, the following is an unentangled state:

$$|\uparrow\downarrow\rangle + |\uparrow\uparrow\rangle = |\uparrow\rangle \otimes (|\downarrow\rangle + |\uparrow\rangle) , \qquad (6.19)$$

while the following is an entangled state:

$$|\downarrow\downarrow\rangle + |\uparrow\uparrow\rangle \neq |\dots\rangle \otimes |\dots\rangle . \qquad (6.20)$$

Entangled states have a variety of uses, including in *Quantum Teleportation*.

Next, let us define density matrices. If the quantum-mechanical wave-function of a system is known (however complex the system might be), its density matrix is defined as:

$$\rho \equiv |\Psi\rangle\langle\Psi| . \qquad (6.21)$$

It can easily be verified that ρ satisfies the following properties:

$$|\rho| \geq 0 , \quad \rho^\dagger = \rho, \quad \rho^2 = \rho . \qquad (6.22)$$

The last property is known as *idempotency*, from which it follows that the eigenvalue p_n of the density matrix can only be 0 or 1. Thus the *Entanglement entropy* or *Von Neumann Entropy*, vanishes :

$$S_{\text{ent}} \equiv -Tr\left(\rho \ln \rho\right) = -\sum_n p_n \ln p_n = 0 . \qquad (6.23)$$

Now, one can take the trace of ρ, only in the subsystem v, to find the *Reduced Density Matrix*, which is still an operator in subsystem u:

$$\rho_u = Tr_v(\rho) = \sum_l \langle v_l|\rho|v_l\rangle = \sum_{i,k,j} d_{ij} d^*_{kj} |u_i\rangle\langle u_k| . \qquad (6.24)$$

For ρ_u, it can be shown that the following properties hold:

$$|\rho_u| \geq 0 , \quad \rho_u^\dagger = \rho_u , \quad \rho_u^2 \neq \rho_u . \qquad (6.25)$$

That is, idempotency no longer holds. As a result, its eigenvalues now satisfy: $0 < p_{n(u)} < 1$, and the entanglement entropy is non-zero:

$$S_u \equiv -Tr_u\left(\rho_u \ln \rho_u\right) = -\sum_n p_{n(u)} \ln p_{n(u)} > 0 . \qquad (6.26)$$

The ignorance resulting from tracing over one part of the system manifests itself as entropy. One important property of reduced density matrices is that if we traced over u instead and found the reduced density matrix ρ_v, then the latter would have the same set of non-zero eigenvalues. Consequently, the entanglement entropies are equal, being a common property of the entangled system:

$$S_v = S_u \,.$$

In the case of black hole entropy, this means that the entanglement entropy obtained by integrating over the degrees of freedom inside the black hole horizon is equal to entropy obtained by integrating over the degrees of freedom outside the horizon.

Having discussed the basics of entanglement, we will now discuss in detail the assumptions and setup for evaluating entanglement in black hole backgrounds.

6.7 Entanglement entropy: Assumptions and setup

We consider a massless scalar field (φ) propagating in an asymptotically flat, four-dimensional Einstenian black hole background given by the Lemaitre line-element[6]:

$$ds^2 = -d\tau^2 + [1 - f(r)]d\xi^2 + r^2 \left[d\theta^2 + \sin^2\theta \, d\phi^2\right] \,, \tag{6.27}$$

where r is the radial coordinate in the Schwarzschild coordinate system and is related to (ξ, τ) by the relation

$$\xi - \tau = \int \frac{dr}{\sqrt{1 - f(r)}} \,, \tag{6.28}$$

$f(r_{\mathrm{H}}) = 0$ where r_{H} is the event horizon. The Hamiltonian of the scalar field propagating in the above line-element is

$$H(\tau) = \frac{1}{2} \int_\tau^\infty d\xi \left[\frac{1}{r^2\sqrt{1 - f(r)}} \Pi_{\ell m}^2 + \frac{r^2}{\sqrt{1 - f(r)}} \left(\partial_\xi \varphi_{\ell m}\right)^2 \right.$$
$$\left. +\ell(\ell + 1)\sqrt{1 - f(r)} \, \varphi_{\ell m}^2 \right] \,, \tag{6.29}$$

where $\varphi_{\ell m}$ is the spherical decomposed field and $\Pi_{\ell m}$ is the canonical conjugate of $\varphi_{\ell m}$ i. e.

$$\varphi_{\ell m}(r) = r \int d\Omega \, Z_{\ell m}(\theta, \phi)\varphi(\vec{r})$$

$$\Pi_{\ell m}(r) = r \int d\Omega \, Z_{\ell m}(\theta, \phi)\Pi(\vec{r}) \,, \tag{6.30}$$

---

[6]The motivation for the choice of scalar fields is given in Ref. [46].

where $Z_{\ell m}(\theta, \phi)$ are the real spherical harmonics. [For simplicity of the notation, we will suppress the subscripts (ℓm).] Although, the above Hamiltonian is time-dependent, there are several advantages of Lemaitre coordinate over the Schwarzschild coordinate [52]: (i) the former is not singular at the horizon (r_{H}) as opposed to the latter, and (ii) ξ (or τ) are space(or, time)-like everywhere while r is space-like only for $r > r_{\mathrm{H}}$.

Having obtained the Hamiltonian, the next step is quantisation. We use Schrödinger representation since it provides a simple and intuitive description of vacuum states for time-dependent Hamiltonian [53]. Formally, we take the basis vector of the state vector space to be the eigenstate of the field operator $\hat{\varphi}(\tau, \xi)$ on a fixed τ hypersurface, with eigenvalues $\varphi(\xi)$ i. e.

$$\hat{\varphi}(\tau, \xi)|\,\varphi(\xi), \tau\,\rangle = \varphi(\xi)|\,\varphi(\xi), \tau\,\rangle \qquad (6.31)$$

The quantum states are explicit functions of time and are represented by wave functionals $\Psi[\varphi(\xi), \tau]$ which satisfy the functional Schrödinger equation:

$$i\frac{\partial \Psi}{\partial \tau} = \int_{\tau}^{\infty} d\xi\, H_{\mathrm{BH}}(\tau)\, \Psi[\varphi(\xi), \tau]. \qquad (6.32)$$

To proceed with the evaluation of S_{ent}:

(i) We assume that the Hamiltonian evolves adiabatically. Technically, this implies that the evolution of the late-time modes leading to Hawking particles are negligible. In the microcanonical ensemble [where the total energy is fixed], this assumption translates to the weak time-dependence of the functional $(\Psi[\varphi(\xi), \tau])$. In the canonical ensemble [where the temperature is fixed], this corresponds to black hole in thermal equilibrium and $\Psi[\varphi(\xi), \tau]$ is approximated as a WKB functional.

(ii) We then obtain ρ_α by tracing the region enclosing the horizon [$\xi \to (r_{\mathrm{H}}, \infty)$] and use Eq. (6.5) to determine S_{ent}[7].

We, now, assume that the above Hamiltonian evolves adiabatically. Technically, this implies that the evolution of the late-time modes leading to Hawking particles are negligible. In the Schrödinger formulation, the above assumption translates to $\Psi[\varphi(\xi), \tau]$ being weakly-dependent on

---

[7]S_{ent} is generally divergent in continuum theories. Therefore usually we assume an ultraviolet cutoff l_{Pl}^2 to regulate the quantum field theory. Below we assume that this is just a technical issue and that we can always have such a regularisation see Refs. [47, 54] and references therein.

time. At a fixed Lemaitre time, Hamiltonian (6.29) reduces to the following flat space-time Hamiltonian [see Appendix (6.12) for details]

$$H = \sum_{\ell m} H_{\ell m} = \sum_{\ell m} \frac{1}{2} \int_0^\infty dr \left\{ \pi_{\ell m}^2(r) + x^2 \left[\frac{\partial}{\partial r} \left(\frac{\varphi_{\ell m}(r)}{r} \right) \right]^2 \right.$$
$$\left. + \frac{\ell(\ell+1)}{r^2} \varphi_{\ell m}^2(r) \right\} . \quad (6.33)$$

The quantum states [cf. Eq. (6.31)] of this Hamiltonian is time-independent and $\Psi[\varphi]$ satisfies the time independent Scrödinger equation

$$\int_0^\infty dr \, H\Psi[\varphi(r)] = E \, \Psi \quad\quad (6.34)$$

6.8   Entanglement entropy: Microcanonical

It is difficult to obtain an analytic expression for the entropy using the Von Neumann definition (6.5). For the field theory, even if we obtain closed-form expression of the density matrix, it is not possible to analytically evaluate the entanglement entropy. Hence, we discretize the Hamiltonian in a spherical lattice of spacing a such that $r \to r_i$; $r_{i+1} - r_i = a$. The ultraviolet cutoff is therefore $M = a^{-1}$. The lattice is large but finite size $L = (N + 1)a$ ($N \gg 1$), with a chosen closed spherical region of radius $R(n + 1/2)a$ inside it. It is this closed region, by tracing over the inside or outside of which one can obtain the reduced density matrix. We demand that the field variables $\varphi_{\ell m}(r) = 0$ for $r \geq L$ so that the infrared cutoff is $\tilde{M} = L^{-1}$. The ultraviolet cutoff is therefore $M = a^{-1}$.

The discretized Hamiltonian is given by

$$H_{\ell m} = \frac{1}{2a} \sum_{j=1}^N \left[\pi_{\ell m,j}^2 + \left[j + \frac{1}{2} \right]^2 \left[\frac{\varphi_{\ell m,j}}{j} - \frac{\varphi_{\ell m,j+1}}{j+1} \right]^2 + \frac{\ell(\ell+1)}{j^2} \varphi_{\ell m,j}^2 \right]$$
$$(6.35)$$

where $\varphi_{\ell m,j} \equiv \varphi_{\ell m}(r_j)$, $\pi_{\ell m} \equiv \pi_{\ell m,j}(r_j)$, which satisfy the canonical commutation relations:

$$[\varphi_{\ell m,j}, \pi_{\ell' m',j'}] = i\delta_{\ell\ell'}\delta_{mm'}\delta_{jj'} . \quad\quad (6.36)$$

Up to an overall factor of a^{-1}, the Hamiltonian $H_{\ell m}$, given by Eq. (6.35), represents the Hamiltonian of $N-$coupled harmonic oscillators (HOs):

$$H_{(N-\text{HO})} = \frac{1}{2} \sum_{i=1}^N p_i^2 + \frac{1}{2} \sum_{i,j=1}^N x_i K_{ij} x_j , \quad\quad (6.37)$$

where the coordinates x_i replace the field variables $\varphi_{\ell m}$, the momenta p_i replace the conjugate momentum variables $\pi_{\ell m}$, and the $N \times N$ matrix K_{ij} $(i, j = 1, \ldots, N)$ represents the potential energy and interaction between the oscillators:

$$
K_{ij} = \frac{1}{i^2} \left[\ell(\ell+1)\, \delta_{ij} + \frac{9}{4}\, \delta_{i1}\delta_{j1} + \left(N - \frac{1}{2}\right)^2 \delta_{iN}\delta_{jN} + \right.
$$
$$
\left. \left\{ \left(i + \frac{1}{2}\right)^2 + \left(i - \frac{1}{2}\right)^2 \right\} \delta_{i,j(i\neq 1,N)} \right]
$$
$$
- \left[\frac{(j + \frac{1}{2})^2}{j(j+1)} \right] \delta_{i,j+1} - \left[\frac{(i + \frac{1}{2})^2}{i(i+1)} \right] \delta_{i,j-1}. \tag{6.38}
$$

The discretization procedure helps in solving the problem semi-analytically, however, to obtain the density matrix, we still need to make a choice of the quantum state. The most general eigen-state of the Hamiltonian (6.37) for the $N-$coupled HOs is given by

$$
\psi(x_1, \ldots, x_N) = \prod_{i=1}^{N} \mathcal{N}_i\, \mathcal{H}_{\nu_i}\left(k_{Di}^{1/4}\, \underline{x}_i\right) \exp\left(-\frac{1}{2}k_{Di}^{1/2}\, \underline{x}_i^2\right), \tag{6.39}
$$

where $\mathcal{N}_i$'s are the normalisation constants given by

$$
\mathcal{N}_i = \frac{k_{Di}^{1/4}}{\pi^{1/4}\,\sqrt{2^{\nu_i}\nu_i!}} \quad, \quad (i = 1, \ldots N), \tag{6.40}
$$

$\underline{x} = Ux$, $(U^T U = I_N)$, $x^T = (x_1, \ldots, x_N)$, $\underline{x}^T = (\underline{x}_1, \ldots, \underline{x}_N)$, $K_D \equiv UKU^T$ is a diagonal matrix with elements k_{Di}, and ν_i $(i = 1 \ldots N)$ are the indices of the Hermite polynomials $(\mathcal{H}_\nu)$. The frequencies are ordered such that $k_{Di} > k_{Dj}$ for $i > j$.

While it is possible to obtain the reduced density matrix and the resultant entropy numerically for the above general state, the results will not provide us any useful physical insight into the problem and especially to identify which states lead to Bekenstein-Hawking entropy and (quantum) corrections. With this motivation, let us make the following choice for the N-particle wave-functional[8]:

$$
\psi_{\mathrm{MS}}(\hat{x}; t) = [c_0\, \psi_{\mathrm{GS}}(\hat{x}; t) + c_1\, \psi_{\mathrm{ES}}(\hat{x}; t)], \tag{6.41}
$$

where ψ_{GS} is the GS wave-function, given by

$$
\psi_{\mathrm{GS}}(x_1, \ldots, x_N) = \prod_{i=1}^{N} \left(\frac{k_{Di}}{\pi}\right)^{1/4} \exp\left(-\frac{1}{2}k_{Di}^{1/2}\underline{x}_i^2\right) \tag{6.42}
$$

---

[8] For other choice of quantum states, see [40, 41, 42, 44]

ψ_{ES} is the ES wave-function, given by

$$\psi_{\mathrm{ES}}(x_1,\ldots,x_N) = \sum_{i=1}^{N}\left(\frac{k_{Di}}{4\pi}\right)^{1/4} \alpha_i \mathcal{H}_1\left(k_{Di}^{1/4}\underline{x}_i\right) \exp\left(-\frac{1}{2}\sum_{j} k_{Dj}^{1/2}\,\underline{x}_j^2\right)$$

$$= \sqrt{2}\left(\alpha^T K_D^{1/2}\underline{x}\right)\psi_{\mathrm{GS}}(x_1,\ldots,x_N), \tag{6.43}$$

$\alpha^T = (\alpha_1,\ldots,\alpha_N)$ are the expansion coefficients, and $\alpha^T\alpha = 1$ so that ψ_{ES} is normalized, $\hat{x} \equiv \{x_1,\cdots,x_n\}$, and $t_j \equiv x_{n+j}$ $(j = 1,\cdots,N-n)$; $t \equiv \{t_1,\cdots,t_{N-n}\} = \{x_{n+1},\cdots,x_N\}$. We assume that c_0 and c_1 are real constants, and ψ_{MS} is normalized so that $c_0^2 + c_1^2 = 1$. Using Eq. (6.43), we can write,

$$\psi_{\mathrm{MS}}(\hat{x};t) = [c_0 + c_1\,f(\hat{x};t)]\,\psi_{\mathrm{GS}}(\hat{x};t),\ f(\hat{x};t) = \sqrt{2}\alpha^T K_D^{1/4}Ux = y^T x, \tag{6.44}$$

where the column vector α includes the expansion coefficients defined in the previous section $[\alpha^T = (\alpha_1,\ldots,\alpha_N) = (1/\sqrt{o})(0,\ldots,0;1,\ldots,1)]$, and y is an N-dimensional column vector y defined as

$$y = \sqrt{2}U^T K_D^{1/4}\alpha = \begin{pmatrix} y_A \\ y_B \end{pmatrix} \tag{6.45}$$

y_A and y_B are n- and $(N-n)$-dimensional column vectors, respectively. The expectation value of energy, $\mathcal{E}$, is given by

$$\mathcal{E} = \langle\psi_{\mathrm{MS}}|H|\psi_{\mathrm{MS}}\rangle = \mathcal{E}_0 + \frac{c_1^2}{o}\sum_{i=N-o+1}^{N} k_{Di}^{1/2}, \tag{6.46}$$

where $\mathcal{E}_0 = \frac{1}{2}\sum_{i=1}^{N} k_{Di}^{1/2}$ is the zero-point energy. The fractional excess of energy over the zero-point energy is therefore given by

$$\frac{\Delta\mathcal{E}}{\mathcal{E}_0} = \frac{\mathcal{E} - \mathcal{E}_0}{\mathcal{E}_0} = \frac{2c_1^2}{o}\left[1 + \frac{\sum_{i=1}^{N-o} k_{Di}^{1/2}}{\sum_{i=N-o+1}^{N} k_{Di}^{1/2}}\right]^{-1}. \tag{6.47}$$

Now, the value of c_1 is between 0 and 1 and as mentioned earlier $k_{Di} > k_{Dj}$ for $i > j$. Therefore, even in the extreme situation $c_1 = 1$, i.e., excited state, with a fairly high amount of excitation $o \sim 50$, the fractional change in energy is at most about $\sim 4\%$. Moreover, since there are o number of terms in the sum in the second term of Eq.(6.46), the excitation energy $(\mathcal{E} - \mathcal{E}_0) \sim 1$ (in units of $1/a$, where a is the lattice spacing)[9].

---

[9]Since the fractional energy is non-zero, we refer to this approach as microcanonical

We now have all the armory to obtain the density matrix and the resultant entropy. The reduced density matrix is obtained by tracing over the first n of the N oscillators:

$$\rho\left(t; t'\right) = \int \prod_{i=1}^{n} dx_i \; \psi_{\mathrm{MS}}(x_1, \ldots, x_n; t) \; \psi_{MS}^{\star}(x_1, \ldots, x_n; t') \quad (6.48)$$

$$= \mathcal{N} \; \exp\left[-\frac{t^T \gamma' t + t'^T \gamma' t'}{2} + t^T \beta' t'\right]$$

where

$$\mathcal{N} = \tilde{\kappa}\sqrt{\frac{|\Omega|}{\pi^{N-n}|A|}} \; \exp\left[-s^T \left(\beta' - \gamma'\right)^T s\right]$$

where β', γ' depend on c_0, c_1, α. (For details, see Ref. [46].).

The $(N - n)$-dimensional constant column vector s is determined from the equation

$$s^T \left(\beta' - \frac{\gamma' + \gamma'^T}{2}\right) = -\kappa_2 \left(y_B - B^T A^{-1} y_A\right), \; \kappa_2 = \frac{c_0 \, c_1}{\tilde{\kappa}} \quad (6.49)$$

where B, A are $n \times n, n \times (N - n)$ matrices, respectively. Note that, for either $c_0 = 0$ or $c_1 = 0$, the constant $\kappa_2 = 0$, whence from the above equation (6.49), we have $s = 0$. It can be verified that the density matrix (6.48) reduces to ground state density matrix when $c_0 = 1, c_1 = 0$.

Computation of the entanglement entropy can been done numerically (using MATLAB), for $N = 300$, $n = 100 - 200$, $o = 30, 40, 50$, and with a precision setting $Pr = 0.01\%$. Figure (6.1) shows the numerical data points of the ratios of the mixed state (MS) entropies — for equal ($c_0 = c_1 = 1/\sqrt{2}$) and high mixings ($c_0 = 1/2, c_1 = \sqrt{3}/2$, with $o = 30, 40, 50$ — to the ground-state entropy (GS) [$c_0 = 1, c_1 = 0$]. The figure also shows the best-fit ratios which follow a simple formula:

$$\frac{S_{\mathrm{MS}}}{S_{GS}} = \tilde{\sigma}_0 + \tilde{\sigma}_1 \left(\frac{A}{a^2}\right)^{-\tilde{\nu}}, \quad (6.50)$$

where the values of the fitting parameters $\tilde{\sigma}_0, \tilde{\sigma}_1$ and $\tilde{\nu}$ are shown in Table 1 for different values of $o = 30, 40, 50$. For all these values of o, the parameter $\tilde{\sigma}_0 \approx 1$ in both MS(Eq) and MS(Hi) cases. The parameter $\tilde{\sigma}_1$ is of the order of 10^3 and increases with increasing excitations. The parameter $\tilde{\nu}$ lies between 1 and 1.25 for the above values of o, and also increases with increasing o. Using the expression for the GS entropy, viz., $S_{GS} = n_0(A/a^2)$, where n_0 is a constant, we can rewrite the above Eq. (6.50) as

$$S_{\mathrm{MS}} = \sigma_0 \left(\frac{A}{a^2}\right) + \sigma_1 \left(\frac{A}{a^2}\right)^{-\nu}, \quad (6.51)$$

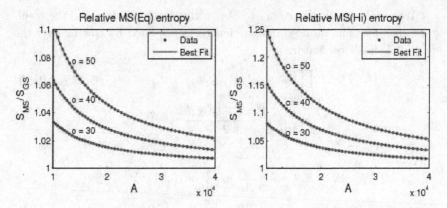

Fig. 6.1 Best fit plots (solid lines) of the relative mixed state entropies ($S_{\mathrm{MS}}/S_{\mathrm{GS}}$) for equal and high mixings versus the area $\mathcal{A}$ (in units of a^2), for $o = 30, 40, 50$. The corresponding data are shown by asterisks.

where $\sigma_0 = n_0 \tilde{\sigma}_0, \sigma_1 = n_0 \tilde{\sigma}_1 \propto c_1$ and $\nu = \tilde{\nu} - 1$. The exponent $-\nu$ lies between 0 and -0.25 for both equal and high mixings with the above values of o. It is instructive to stress the implications of the above result:

(i) For the pure vacuum wave-functional, $c_1 = 0$ and S_{MS} is identical to Bekenstein-Hawking entropy. This clearly shows that the entanglement entropy of ground state leads to the area law and the excited states contribute to the power-law corrections for black-holes in Einstenian gravity.

(ii) For large black-holes, power-law correction falls off rapidly and we recover S_{BH}. However, for the small black-holes, the second term dominates and black hole entropy is no more proportional to area. Physical interpretation of this result is immediately apparent. In the large black hole (or low-energy) limit, it is difficult to excite the modes and hence, the ground state modes contribute significantly to S_{MS}. However, in the small black hole (or high-energy) limit, larger number of field modes can be excited and hence they contribute significantly to S_{MS}.

(iii) The corrections to the Bekenstein-Hawking area law derived here are obtained for Einstenian gravity. This has to be contrasted with other approaches like Noether charge where the sub-leading correction to area-law appear due to the presence of higher-derivative terms. As mentioned in Sec. (6.4), the corrections in these approaches follow directly from the gravity action. However, in our case, the corrections are quantum corrections and do not have (semi-)classical correspondence.

6.9 Entanglement entropy: Canonical

In the microcanonical ensemble approach, we need to resort to numerical computations to obtain entanglement entropy. To confirm the numerical results and also gain insight about the dependence of the entropy with the kinematical properties of black hole horizon r_H, we now obtain entanglement entropy in canonical ensemble approach.

In the adiabatic limit, the ansatz for the wave-functional is

$$\Psi[\varphi(\xi), \tau] = P[\varphi] \exp\left[\frac{i}{\hbar} S[\varphi(\xi, \tau)]\right] \tag{6.52}$$

where S is the Hamilton-Jacobi functional which satisfy

$$\frac{\partial S}{\partial \tau} + H\left(\varphi_{\ell m}, \frac{\delta S}{\delta \varphi_{\ell m}}\right) = 0, \tag{6.53}$$

H is (6.29) and $P[\varphi]$ is the 1-loop term. The density of states can then be written as

$$\Gamma(E) = \frac{1}{\pi} \int_{r_H + h_c}^{L} dr \int_0^{\ell_{max}} d\ell \, (2\ell + 1) \frac{dS}{dr} \tag{6.54}$$

$S(r)$ can not be known for a general $f(r)$ [cf. Eq. (6.27)]. Using the fact that the density of states grow close to the horizon, we Taylor expand $f(r)$ near the horizon i. e.

$$f(r) = f'(r_H)(r - r_H) + \left(\frac{f''(r_H)}{2}\right)(r - r_H)^2 \tag{6.55}$$
$$+ \left(\frac{f'''(r_H)}{6}\right)(r - r_H)^3 + \dots$$

Substituting the above expressions, in the expression for free-energy and entropy:

$$F(\beta) = -\int_0^\infty dE \, \frac{\Gamma(E)}{\exp(\beta E) - 1} \qquad S_c(\beta) = \beta^2 \left(\frac{\partial F}{\partial \beta}\right) \tag{6.56}$$

where $\beta = 1/T$, leads to the master equation (for more details, see, Ref. [47])

$$S_c(\beta) = S_{BH} + \mathcal{F}^{(4D)}(\mathcal{A}_H) \log\left(\frac{\mathcal{A}_H}{l_{Pl}^2}\right) \tag{6.57}$$

where

$$\mathcal{F}^{(4D)}(\mathcal{A}_H) = -\left(\frac{1}{60}\right) f''(r_H) \, r_H^2 + \left(\frac{1}{10}\right) \kappa \, r_H \,. \tag{6.58}$$

We would like to stress a few points regarding this result:

(1) This is a master equation and gives the entropy corresponding to a general spherically symmetric black hole space-time in Einstenian gravity. The sub-leading corrections depend only on the kinematical properties of black hole i. e. surface gravity and second derivative of metric function. (It should be noted that this form is unique for all orders of the WKB approximation and does not depend on third and higher order derivatives of the metric [47].)

(2) $\mathcal{F}$ is a constant — and hence, sub-leading corrections are purely logarithmic — only, if $\kappa \propto r_{\mathrm{H}}^{-1}$ and $f''(r_{\mathrm{H}}) \propto r_{\mathrm{H}}^{-2}$. This uniquely corresponds to Schwarzschild space-time. For any other black hole space-times in Einstenian gravity like, for instance, Reissner-Nördstrom, Schwarzschild-(Anti)de Sitter, we have

$$S^{\mathrm{c}}_{\mathrm{ent}} = \mathcal{S}_{\mathrm{BH}} - \frac{\pi^{1/2}}{15} \left(\frac{l_{\mathrm{Pl}}^2}{\mathcal{A}_{\mathrm{H}}} \right)^{-1/2} \log \left(\frac{\mathcal{A}_{\mathrm{H}}}{l_{\mathrm{Pl}}^2} \right) + \text{Higher contributions}$$

(6.59)

(3) As in the microcanonical ensemble, (a) in the large black hole limit the power-law corrections fall off rapidly and we recover S_{BH} (b) in the small black hole limit, the second term dominates and the black hole entropy is not proportional to area.

6.10 Conclusions and discussion

In the absence of a consistent quantum theory of gravity, the best strategy is to slowly build a coherent picture and hope to understand — and, in due course, solve — some of the problems of black hole thermodynamics. In this article, we have raised two questions which are relevant for any approach: (i) Using simple techniques, can we know whether an approach to black hole entropy contains semi-classical or quantum degrees of freedom? (ii) Is it possible to identify generic quantum corrections to Bekenstein-Hawking entropy?

Using naive-dimensional analysis, we show that it is possible to differentiate between different approaches. We have shown that this observation suggests that if any approach to black hole entropy predicts the same dimensional form as that of the classical action of gravity, then this approach *only* provides the semi-classical part of the theory and does not provide information about the quantum structure of gravity. This might seem a strong assertion, however, it would be an even stronger claim if one says that the quantum corrections to the semi-classical black hole entropy follow the same dimensionality of the classical action. This also implies that the recent results from string theory using the attractor mechanism for extremal black-holes also provides information about the semi-classical aspect of gravity and does not necessarily provide information about quantum aspect of gravity (for a recent review, see Ref. [30]).

Quantum entanglement as a source of black hole entropy stands out for its simplicity and generality. The results discussed in this review highlight the nontrivial, and somewhat counter-intuitive, facets of quantum entanglement and its role as the source of black hole entropy. More precisely, assuming the modes evolve adiabatically, we have shown that:

- Entanglement leads to generic power-law corrections to the area law
- The quantum degrees of freedom that lead to S_{BH} and sub-leading corrections are different.
- It is possible to identify the quantum degrees of freedom that contribute to the area law and the sub-leading corrections.

From the above conclusions, one can ask the reverse question: Starting from the black hole entropy (with quantum corrections), can one predict the effective quantum corrections to the classical gravity action? Assuming that the naive dimensional discussed in sec. (6.4) continue to hold for the corrections arising in entanglement approach, then it is possible to obtain effective quantum corrections to the Einstein-Hilbert (6.6).

This leads to the following interesting questions: What kind of black hole solutions can we obtain from this effective gravity action? Are the space-times geodesically complete or incomplete? Do they satisfy the laws of black hole thermodynamics [18, 19, 3]? We hope to report on these in future.

6.11   Acknowledgments

The author wishes to thank Saurya Das, Sudipta Sarkar, L. Sriramkumar and Sourav Sur for discussions.

6.12   Appendix

In this appendix, we find the expression for the Hamiltonian of a scalar field propagating in a static spherically symmetric space-time and show that for a particular time slicing this Hamiltonian reduces to that of a scalar field in flat space-time.

Let us consider the line-element for a general four-dimensional spherically symmetric space-time:

$$ds^2 = -A(\tau,\xi)\,d\tau^2 + \frac{d\xi^2}{B(\tau,\xi)} + \rho^2(\tau,\xi)\left(d\theta^2 + \sin^2\theta d\phi^2\right), \qquad (6.60)$$

where A, B, ρ are continuous, differentiable functions of (τ,ξ). The action for the scalar field φ propagating in this space-time is given by

$$S = -\frac{1}{2}\int d^4x\,\sqrt{-g}\,g^{\mu\nu}\,\partial_\mu\varphi\,\partial_\nu\varphi \qquad (6.61)$$

$$= -\frac{1}{2}\sum_{\ell m}\int d\tau d\xi\Big[-\frac{\rho^2}{\sqrt{AB}}(\partial_\tau\varphi_{\ell m})^2 + \sqrt{AB}\rho^2\,(\partial_\xi\varphi_{\ell m})^2$$

$$+ \ell(\ell+1)\sqrt{\frac{A}{B}}\,\varphi_{\ell m}^2\Big].$$

where we have decomposed φ in terms of the real spherical harmonics $(Z_{\ell m}(\theta,\phi))$:

$$\varphi(x^\mu) = \sum_{\ell m}\varphi_{\ell m}(\tau,\xi)Z_{\ell m}(\theta,\phi). \qquad (6.62)$$

Following the standard rules, the canonical momenta and Hamiltonian of the field are given by

$$H_{\ell m}(\tau) = \frac{1}{2}\int_\tau^\infty d\xi\Big[\frac{\sqrt{AB}}{\rho^2}\Pi_{\ell m}^2 + \sqrt{AB}\,\rho^2(\partial_\xi\varphi_{\ell m})^2$$

$$+\ell(\ell+1)\sqrt{\frac{A}{B}}\,\varphi_{\ell m}^2\Big] \qquad (6.63)$$

$$\Pi_{\ell m} = \frac{\partial\mathcal{L}}{\partial(\partial_\tau\varphi_{\ell m})} = \frac{\rho^2}{\sqrt{AB}}\,\partial_\tau\varphi_{\ell m}\,, H = \sum_{\ell m}H_{\ell m}. \qquad (6.64)$$

The canonical variables $(\varphi_{\ell m}, \Pi_{\ell m})$ satisfy the Poisson brackets

$$\{\varphi_{\ell m}(\tau,\xi), \Pi_{\ell m}(\tau,\xi')\} = \delta(\xi - \xi') \qquad (6.65)$$

$$\{\varphi_{\ell m}(\tau,\xi), \varphi_{\ell m}(\tau,\xi')\} = 0 = \{\Pi_{\ell m}(\tau,\xi), \Pi_{\ell m}(\tau,\xi')\}.$$

In the time-dependent Lemaitre coordinates [52, 55] the metric components of the line-element (6.60) are given by

$$A(\tau,\xi) = 1 \quad ; \quad B(\tau,\xi) = \frac{1}{1 - f(r)} \quad ; \quad \rho(\tau,\xi) = r\,, \qquad (6.66)$$

where $r = r(\tau, \xi)$. The line-element in the Lemaitre coordinates is related to that in the time-independent Schwarzschild coordinates, viz.,

$$ds^2 = -f(r)dt^2 + \frac{dr}{f(r)} + r^2 \left(d\theta^2 + \sin^2\theta d\phi^2\right) \quad ; \quad f(r = r_h) = 0 \quad (6.67)$$

by the following transformations [55]:

$$\tau = t \pm \int dr \frac{\sqrt{1 - f(r)}}{f(r)} \quad ; \quad \xi = t + \int dr \frac{[1 - f(r)]^{-1/2}}{f(r)} \quad . \quad (6.68)$$

Unlike the line-element in Schwarzschild coordinates, the line-element in Lemaitre coordinates is not singular at the horizon r_h. Moreover, the coordinate ξ (or, τ) is space(or, time)-like everywhere, whereas r(or, t) is space(or, time)-like only for $r > r_h$.

In Lemaitre coordinates the general Hamiltonian (6.63) takes the form

$$H_{\ell m}(\tau) = \frac{1}{2} \int_\tau^\infty d\xi \left[\frac{1}{r^2\sqrt{1 - f(r)}} \Pi_{\ell m}^2 + \frac{r^2}{\sqrt{1 - f(r)}} (\partial_\xi \varphi_{\ell m})^2 \right. \quad (6.69)$$
$$\left. + \ell(\ell + 1)\sqrt{1 - f(r)}\, \varphi_{\ell m}^2 \right] ,$$

which depends explicitly on the Lemaitre time.

Choosing now a fixed Lemaitre time ($\tau = \tau_0 = 0$, say), (6.68) lead to:

$$\frac{d\xi}{dr} = \frac{1}{\sqrt{1 - f(r)}} . \quad (6.70)$$

If we set $d\theta = d\phi = 0$, then for the fixed Lemaitre time τ_0 it follows that $ds^2 = d\xi^2/B(\tau_0, \xi) = dr^2$, i.e., the covariant cut-off is $|ds| = dr$. Substituting the above relation (6.70) in the Hamiltonian (6.69) we get

$$H_{\ell m}(0) = \frac{1}{2} \int_0^\infty dr \left[\frac{\Pi_{\ell m}^2\, r^{-2}}{1 - f(r)} + r^2 (\partial_r \varphi_{\ell m})^2 + \ell(\ell + 1)\, \varphi_{\ell m}^2 \right] , \quad (6.71)$$

where the variables $(\varphi_{\ell m}, \Pi_{\ell m})$ satisfy the relation:

$$\{\varphi_{\ell m}(r), \Pi_{\ell m}(r')\} = \sqrt{1 - f(r)}\delta(r - r'). \quad (6.72)$$

Performing the following canonical transformations

$$\Pi_{\ell m} \to r\sqrt{1 - f(r)}\, \Pi_{\ell m} \; ; \; \varphi_{\ell m} \to \frac{\varphi_{\ell m}}{r} \quad (6.73)$$

the full Hamiltonian reduces to that of a free scalar field propagating in flat space-time [56]

$$H = \sum_{\ell m} \frac{1}{2} \int_0^\infty dr \left\{ \pi_{\ell m}^2(r) + r^2 \left[\frac{\partial}{\partial r} \left(\frac{\varphi_{\ell m}(r)}{r} \right) \right]^2 + \frac{\ell(\ell + 1)}{r^2}\, \varphi_{\ell m}^2(r) \right\} (6.74)$$

This happens for *any* fixed value of the Lemaitre time τ, provided the scalar field is traced over either the region $r \in (0, r_h]$ or the region $r \in [r_h, \infty)$. Note that the black hole singularity can be entirely avoided for the latter choice, and for evaluating time-independent quantities such as entropy, it suffices to use the above Hamiltonian.

References

[1] T. Padmanabhan, "Acceptable density perturbations from inflation due to quantum gravitational damping," *Phys. Rev. Lett.* **60** (1988) 2229.

[2] T. Padmanabhan, "Gravity as elasticity of spacetime: A paradigm to understand horizon thermodynamics and cosmological constant," *Int. J. Mod. Phys.* **D13** (2004) 2293–2298, arXiv:gr-qc/0408051.

[3] T. Padmanabhan, "Gravity and the thermodynamics of horizons," *Phys. Rept.* **406** (2005) 49–125, arXiv:gr-qc/0311036.

[4] T. Padmanabhan, "Holographic Gravity and the Surface term in the Einstein- Hilbert Action," *Braz. J. Phys.* **35** (2005) 362–372, arXiv:gr-qc/0412068.

[5] T. Padmanabhan, "Gravity as an emergent phenomenon: A conceptual description," *AIP Conf. Proc.* **939** (2007) 114–123, arXiv:0706.1654 [gr-qc].

[6] T. Padmanabhan, "Emergent gravity and Dark Energy," arXiv:0802.1798 [gr-qc].

[7] T. Padmanabhan, "From gravitons to gravity: Myths and reality," *Int. J. Mod. Phys.* **D17** (2008) 367–398, arXiv:gr-qc/0409089.

[8] T. Padmanabhan, "A Physical Interpretation of Gravitational Field Equations," arXiv:0911.1403 [gr-qc].

[9] A. Wehrl, "General properties of entropy," *Rev. Mod. Phys.* **50** (1978) 221–260.

[10] S. W. Hawking, "Black hole explosions," *Nature* **248** (1974) 30–31.

[11] J. von Neumann, *Mathematical Foundations of Quantum Mechanics.* Princeton University press, 1996.

[12] J. A. Wheeler, "Sakharov revisited: It from bit," in *Sakharov Memorial Lecture on Physics 2*, L. Keldysh and V. Feinberg, eds. Nova, 1992. In *Moscow 1991, Proceedings, Sakharov memorial lectures in physics, vol. 2* 751-769.

[13] R. Brustein, M. B. Einhorn, and A. Yarom, "Entanglement interpretation of black hole entropy in string theory," *JHEP* **01** (2006) 098, arXiv:hep-th/0508217.

[14] T. Nishioka, S. Ryu, and T. Takayanagi, "Holographic Entanglement Entropy: An Overview," *J. Phys.* **A42** (2009) 504008, arXiv:0905.0932 [hep-th].

[15] R. M. Wald, "The thermodynamics of black holes," *Liv. Rev. Rela.* **4** (2001) 6, arXiv:gr-qc/9912119. http://www.livingreviews.org/

lrr-2001-6.

[16] V. P. Frolov and I. D. Novikov, *Black hole physics: Basic concepts and new developments*. Kluwer Academic, 1998.

[17] A. Manohar and H. Georgi, "Chiral Quarks and the Nonrelativistic Quark Model," *Nucl. Phys.* **B234** (1984) 189.

[18] J. D. Bekenstein, "Black holes and entropy," *Phys. Rev.* **D7** (1973) 2333–2346.

[19] S. W. Hawking, "Black holes and thermodynamics," *Phys. Rev.* **D13** (1976) 191–197.

[20] D. Lovelock, "The Einstein tensor and its generalizations," *J. Math. Phys.* **12** (1971) 498–501.

[21] D. Kothawala, T. Padmanabhan, and S. Sarkar, "Is gravitational entropy quantized ?," *Phys. Rev.* **D78** (2008) 104018, arXiv:0807.1481 [gr-qc].

[22] D. Kothawala and T. Padmanabhan, "Thermodynamic structure of Lanczos-Lovelock field equations from near-horizon symmetries," *Phys. Rev.* **D79** (2009) 104020, arXiv:0904.0215 [gr-qc].

[23] R. Wald, "Black hole entropy in the Noether charge," *Phys. Rev.* **D48** (1993) R3427–R3431, gr-qc/9307038.

[24] R. C. Myers and M. J. Perry, "Black Holes in Higher Dimensional Space-Times," *Annals Phys.* **172** (1986) 304.

[25] R. C. Myers and J. Z. Simon, "Black hole evaporation and higher derivative gravity," *Gen. Rel. Grav.* **21** (1989) 761–766.

[26] R. C. Myers and J. Z. Simon, "Black Hole Thermodynamics in Lovelock Gravity," *Phys. Rev.* **D38** (1988) 2434–2444.

[27] T. Jacobson and R. C. Myers, "Black hole entropy and higher curvature interactions," *Phys. Rev. Lett.* **70** (1993) 3684–3687, arXiv:hep-th/9305016.

[28] T. Jacobson, G. Kang, and R. C. Myers, "Increase of black hole entropy in higher curvature gravity," *Phys. Rev.* **D52** (1995) 3518–3528, arXiv:gr-qc/9503020.

[29] A. Paranjape, S. Sarkar, and T. Padmanabhan, "Thermodynamic route to field equations in Lancos-Lovelock gravity," *Phys. Rev.* **D74** (2006) 104015, arXiv:hep-th/0607240.

[30] A. Sen, "Black Hole Entropy Function, Attractors and Precision Counting of Microstates," *Gen. Rel. Grav.* **40** (2008) 2249–2431, arXiv:0708.1270 [hep-th].

[31] S. Carlip, "Logarithmic corrections to black hole entropy from the Cardy formula," *Class. Quant. Grav.* **17** (2000) 4175–4186, arXiv:

gr-qc/0005017.

[32] R. K. Kaul and P. Majumdar, "Logarithmic correction to the Bekenstein-Hawking entropy," *Phys. Rev. Lett.* **84** (2000) 5255–5257, gr-qc/0002040.

[33] R. K. Pathria, *Statistical mechanics*. Butterworth-Heinemann, Oxford ; Boston :, 2nd ed. ed., 1996.

[34] R. Reid, J. Prausnitz, and B. Poling, *The properties of gases and liquids*. McGraw-Hill, 4th ed., 1987.

[35] A. Strominger and C. Vafa, "Microscopic Origin of the Bekenstein-Hawking Entropy," *Phys. Lett.* **B379** (1996) 99–104, hep-th/9601029.

[36] A. Ashtekar, J. Baez, A. Corichi, and K. Krasnov, "Quantum geometry and black hole entropy," *Phys. Rev. Lett.* **80** (1998) 904–907, gr-qc/9710007.

[37] S. Carlip, "Entropy from conformal field theory at Killing horizons," *Class. Quant. Grav.* **16** (1999) 3327–3348, arXiv:gr-qc/9906126.

[38] G. 't Hooft, "On the quantum structure of a black hole," *Nucl. Phys.* **B256** (1985) 727.

[39] L. Bombelli, R. K. Koul, J.-H. Lee, and R. D. Sorkin, "A Quantum Source of Entropy for Black Holes," *Phys. Rev.* **D34** (1986) 373.

[40] M. Srednicki, "Entropy and area," *Phys. Rev. Lett.* **71** (1993) 666–669, hep-th/9303048.

[41] M. Ahmadi, S. Das, and S. Shankaranarayanan, "Is entanglement entropy proportional to area?," *Can. J. Phys.* **84** (2006) 493–499, arXiv:hep-th/0507228.

[42] S. Das and S. Shankaranarayanan, "How robust is the entanglement entropy - area relation?," *Phys. Rev.* **D73** (2006) 121701, arXiv:gr-qc/0511066.

[43] S. Das and S. Shankaranarayanan, "Where are the black hole entropy degrees of freedom?," *Class. Quant. Grav.* **24** (2007) 5299–5306, arXiv:gr-qc/0703082.

[44] S. Das and S. Shankaranarayanan, "Entanglement as a source of black hole entropy," *J. Phys. Conf. Ser.* **68** (2007) 012015, arXiv:gr-qc/0610022.

[45] S. Das, S. Shankaranarayanan, and S. Sur, "Where are the degrees of freedom responsible for black hole entropy?," *Can. J. Phys.* **86** (2008) 653–658, arXiv:0708.2098 [gr-qc].

[46] S. Das, S. Shankaranarayanan, and S. Sur, "Power-law corrections to entanglement entropy of black holes," *Phys. Rev.* **D77** (2008) 064013, arXiv:0705.2070 [gr-qc].

[47] S. Sarkar, S. Shankaranarayanan, and L. Sriramkumar, "Sub-leading contributions to the black hole entropy in the brick wall approach," *Phys. Rev.* **D78** (2008) 024003, arXiv:0710.2013 [gr-qc].

[48] S. Shankaranarayanan, "Do subleading corrections to Bekenstein-Hawking entropy hold the key to quantum gravity?," *Mod. Phys. Lett.* **A23** (2008) 1975–1980, arXiv:0805.4531 [gr-qc].

[49] S. Das, S. Shankaranarayanan, and S. Sur, "Black hole entropy from entanglement: A review," arXiv:0806.0402 [gr-qc].

[50] S. Das, P. Majumdar, and R. K. Bhaduri, "General logarithmic corrections to black hole entropy," *Class. Quant. Grav.* **19** (2002) 2355–2368, arXiv:hep-th/0111001.

[51] P. Tommasini, E. Timmermans, and A. F. R. de Toledo Piza, "The Hydrogen Atom as an Entangled Electron-Proton System," arXiv:quant-ph/9709052.

[52] L. Landau and E. M. Lifshitz, *Course of Theoretical Physics, Vol. 2, The Classical theory of fields.* Pergamon Press, London, 1975.

[53] B. Hatfield, *Quantum field theory of point particles and strings.* Addison-Wesley (Frontiers in Physics), 1992. Redwood City, USA: Addison-Wesley (1992) 734 p. (Frontiers in physics, 75).

[54] R. Brout, "Entanglement and Thermodynamics of Black Hole Entropy," *Int. J. Mod. Phys.* **D17** (2009) 2549–2553, arXiv:0802.1588 [gr-qc].

[55] S. Shankaranarayanan, "Temperature and entropy of Schwarzschild-de Sitter space- time," *Phys. Rev.* **D67** (2003) 084026, arXiv:gr-qc/0301090.

[56] K. Melnikov and M. Weinstein, "On unitary evolution of a massless scalar field in a Schwarzschild background: Hawking radiation and the information paradox," *Int. J. Mod. Phys.* **D13** (2004) 1595–1636, arXiv:hep-th/0205223.

Chapter 7

Quantum measurement and quantum gravity: many worlds or collapse of the wave function?[1]

T. P. Singh

Tata Institute of Fundamental Research,
Homi Bhabha Road, Mumbai 400 005, India.
E-mail: *tpsingh@tifr.res.in*

Abstract: At present, there are two possible, and equally plausible, explanations for the physics of quantum measurement. The first explanation, known as the many-worlds interpretation, does not require any modification of quantum mechanics, and asserts that at the time of measurement the Universe splits into many branches, one branch for every possible alternative. The various branches do not interfere with each other because of decoherence, thus providing a picture broadly consistent with the observed Universe. The second explanation, which requires quantum mechanics to be modified from its presently known form, is that at the time of measurement the wave-function collapses into one of the possible alternatives. The two explanations are mutually exclusive, and up until now, no theoretical reasoning has been put forward to choose one explanation over the other. In this article, we provide an argument which implies that the collapse interpretation is favored over the many-worlds interpretation. Our starting point is the assertion (which we justify) that there ought to exist a reformulation of quantum mechanics which does not refer to a classical spacetime manifold. The need for such a reformulation implies that quantum theory becomes non-linear on the Planck mass/energy scale. Standard linear quantum mechanics is an approximation to this non-linear theory, valid at energy scales much smaller than the Planck scale. Using ideas based on noncommutative differential geometry, we develop such a reformulation and derive a non-linear Schrödinger equation, which

---

[1]This article originally appeared in J. Phys.: Conf. Ser. **174**, 012024 (2009). © 2009 IOP Publishing Ltd. Reprinted with permission.

can explain collapse of the wave-function. We also obtain an expression for the lifetime of a quantum superposition. We suggest ideas for an experimental test of this model.

7.1 The quantum measurement problem

Suppose that in a quantum measurement one measures an observable $\hat{O}$ of a quantum system, and suppose that prior to the measurement the system is in a state $|\psi>$ which can be expanded in a basis of orthonormal eigenstates $|\psi_n>$ of the observable $\hat{O}$ as

$$|\psi> = \sum_n a_n |\psi_n> . \qquad (7.1)$$

It is then known from experiment that, if the selected states of the measuring apparatus are in one-to-one correspondence with the basis $|\psi_n>$, then after the measurement the quantum system is found to be in one of the eigenstates, say $|\psi_n>$. Repeated measurements on identical copies of the quantum system show that the system is found to be in one or the other eigenstates $|\psi_n>$, with the probability to be found in state $|\psi_n>$ being given by $|a_n|^2$ (the Born probability rule).

The transition $|\psi> \rightarrow |\psi_n>$ that takes place during a quantum measurement cannot be described by the Schrödinger equation. This is of course because the Schrödinger equation is linear, and Schrödinger evolution will preserve the superposition expressed in Eqn. (7.1). The observed transition $|\psi> \rightarrow |\psi_n>$, on the other hand, breaks linear superposition. The quantum measurement problem can be stated as follows: what is the correct physical description of this measurement process and of the observed result? This description should explain why the transition takes place in the first place, and why it obeys the Born probability rule. There are two possible explanations, which we elaborate on below.

7.1.1 *First explanation: The many-worlds interpretation*

According to the many-worlds interpretation of quantum mechanics, (originally due to Everett [1]), despite appearances, the transition $|\psi> \rightarrow |\psi_n>$ in fact does not take place at all during a quantum measurement, and superposition continues to be preserved after the measurement has taken place. Rather, it is assumed that during a quantum measurement the Universe splits into many branches, with a particular outcome, say $|\psi_n>$, being realized in our branch of the Universe. Here, the term Universe is meant to refer also to the measuring apparatus, and the observer as well. It has also been asserted that the many-worlds interpretation is consistent with the Born probability rule [2]. (For another discussion on the probability rule in the many-worlds picture see [3].) A recent account of the many-worlds interpretation, including recent developments in its understanding, can be found in [4].

The different branches of the Universe do not interfere with each other because of the phenomenon of decoherence. The process of decoherence, which has been experimentally observed [5], destroys the interference amongst various alternatives in the quantum state of a macroscopic system consisting of a superposition of various alternatives [6]. Thus in the many-worlds picture the various branches of the Universe continue to remain superposed, as required by Schrödinger evolution, but do not interfere with each other, as a consequence of decoherence. In this picture, quantum mechanics does not have to be modified in order to explain a quantum measurement. This has been called 'economy of assumptions, and extravagance of Universes'.

The many-worlds interpretation may appear counter-intuitive, but it is difficult to find logical inconsistencies in the interpretation. The picture is in fact attractive because it can explain quantum measurement without having to change the laws of quantum mechanics. Its shortcoming perhaps is that it does not appear to be experimentally falsifiable (unless experimental proof can be found for the 'collapse' interpretation discussed next). Also, if the Universe splits into many branches, it is not clear how the obscure issue of 'splitting of the consciousness of an observer into many branches' is to be understood. Neither of these shortcomings however can by themselves rule out the possibility that the many-worlds interpretation could be the correct explanation of quantum measurement.

7.1.2 *Second explanation: Collapse of the wave-function*

The second explanation is that the transition $|\psi> \to |\psi_n>$ does indeed take place, and the Universe does not split into many branches. It is assumed that there is only one branch to the Universe, the one that we directly observe, and live in. The two explanations (many-worlds and collapse) are clearly mutually exclusive : one, and only one, out of the two explanations must be correct.

In order to explain quantum measurement by invoking collapse of the wave-function, quantum mechanics and the Schrödinger equation must be modified. Quantum mechanics as we presently know it can only be an approximation to a more general theory, with the more general theory having the capacity to explain wave-function collapse. This has been called 'economy of Universes, and extravagance of assumptions'.

For instance, it may be possible to explain quantum measurement by generalizing the Schrödinger equation to a non-linear Schrödinger equation. The non-linearity is assumed to become important in the measurement domain, but is negligible in the microscopic domain. In principle, the presence of the non-linearity can result in breakdown of superposition, driving the quantum system to one particular alternative, in a manner consistent with the Born probability rule. This particular approach to collapse of the wave-function will be the focus of the present paper. Other models of collapse are briefly reviewed in Section 7.4. Objections against a non-linear quantum mechanics (such as superluminality) are briefly addressed in the Discussion section.

It is only fair to say that as of now, there is no universally accepted theory for collapse of the wave-function, supported by experiment. Neither is there any experimental evidence that quantum mechanics has to be modified from its present form. Needless to say, this situation could change in the future.

7.1.3 *Goal of the present paper*

Up until now, there has been no experimental or theoretical motivation to favor the many-worlds interpretation of quantum measurement over the collapse interpretation, or vice-versa. Critics of the many-worlds interpretation could say its unfalsifiable, while adherents of this interpretation consider it an advantage that it requires no changes in the existing formulation of quantum theory. Proponents of the collapse model find it unphysical that unobservable parallel Universes are invoked so as to protect the prevailing structure of quantum theory, whereas critics label the collapse models as *ad hoc*; having been invented for the sole purpose of explaining quantum measurement, and not embedded in a broader theoretical framework.

The purpose of the present paper is to argue that there are additional theoretical reasons, hitherto unemphasized, which suggest that the collapse explanation is favored over the many-worlds interpretation. Our starting point, which we elaborate on in detail in Section 7.3, is that the notion of time which is used to describe time evolution in quantum theory is a classical notion. There ought to exist a reformulation of quantum mechanics which does not make a reference to this external classical notion of time. Furthermore, under appropriate circumstances, i.e. as and when one chooses to make reference to (a possibly available) external classical time, this reformulation should become equivalent to standard quantum mechanics.

The central thesis of our paper is that the requirement that there be such a reformulation of quantum mechanics leads to the conclusion that standard quantum theory is a limiting case of a more general, non-linear, quantum theory. The non-linearity becomes important in the measurement domain, and could cause the collapse of the wave-function during a quantum measurement. This is why we say that the collapse based explanation is favored over the many-worlds interpretation. We arrive at this conclusion, not in an *ad hoc* fashion, but by addressing an entirely different incompleteness in the existing formulation of quantum theory, namely, its undesirable reference to an external classical time. Thus our conclusion is that removing the notion of classical time from quantum mechanics has an important and significant byproduct - one may be able to explain quantum measurement as a consequence of wave-function collapse. The requirement that the aforementioned reformulation of quantum mechanics should exist is unavoidable; and non-linearity in quantum mechanics is its inevitable consequence. The many-worlds interpretation is thus disfavored, not because it might appear to be counter-intuitive and discomforting, but because of compelling theoretical reasoning which extends outside and beyond the current formulation of quantum mechanics.

In Section 7.3 we will propose a tentative reformulation of quantum mechanics, borrowing ideas from noncommutative geometry. We will arrive at a non-linear Schrödinger equation which generalizes the standard linear Schrödinger equation, and which can explain at least some significant aspects of the process of quantum measurement.

Before we present this reformulation, we will review in the next Section an illustrative toy-model for collapse induced by non-linearity, which is due to Grigorenko [7]. This model will be of help to us in understanding the relation between wave-function collapse and the non-linear equation we arrive at in Section 7.3.

[A few other significant interpretations of quantum mechanics, not discussed in the present paper, have also been proposed in the literature. These include the works of Bohm [8] (and its generalization to non-local hidden variable theories), and those of Gell-Mann and Hartle [9], and Omnes [10].

These works present a different formulation of standard quantum mechanics, without implying that the experimental predictions of the new formulation differ from that of the standard theory.]

7.2   A toy model for non-linear quantum mechanics and collapse of the wave-function

7.2.1   *Introduction*

We propose a non-linear Schrödinger equation which during a quantum measurement can dynamically induce the transition $|\psi> \rightarrow |\psi_n>$ with a probability $|a_n|^2$, when the initial state of the quantum system is given by Eqn. (7.1). Since such a non-linear equation must not violate what is known about quantum mechanics from experiments (including stringent bounds on non-linearity in the microscopic domain), it must satisfy several conditions, which we enumerate below:

- The non-linearity must become significant only during the onset of a quantum measurement, and not before that.

- The non-linear equation must reduce to the standard linear Schrödinger equation when applied to the special case of microscopic systems.

- The non-linear equation must reduce to the standard equation of motion of classical mechanics when applied to the special case of macroscopic systems.

- The Hamiltonian must have a non-Hermitean part, which is also non-linear, so that the initial superposition of states can decay into one of the alternatives. However, the action of the full Hamiltonian on the states must be norm-preserving.

- Since in this scenario the outcome of a quantum measurement is deterministic but random, the non-linear equation must contain one or more random variables. One or the other outcome of a measurement is realized, in consistency with the Born rule, depending on the relative values of the random variables.

- The predictions of the non-linear equation must be experimentally testable, and must not contradict the experimentally verified features of standard quantum mechanics.

- The non-linear equation must not be *ad hoc*, but must instead be a consequence of 'some other requirement'.   In other words, there must be good theoretical reasons, independent of quantum measurement, such as those outlined in Section 7.1.3, for such a non-linear generalization of quantum mechanics.

In this Section, we will work with a non-linear Schrödinger equation which belongs to the following general class of norm-preserving non-linear Schrödinger equations:

$$i\hbar d|\psi > /dt = H|\psi > +(1 - P_\psi)U|\psi > . \qquad (7.2)$$

Here, H is the Hermitian part of the Hamiltonian, as in standard quantum mechanics. $(1 - P_\psi)U$ is the non-Hermitian part, $P_\psi = |\psi><\psi|$ is the projection operator, and U is an arbitrary nonlinear operator. This equation has been discussed by [11] and by [7]. We will make the crucial assumption (to be justified later, in Section 7.3) that the non-Hermitean part of the Hamiltonian becomes significant only when the mass of the system becomes comparable to or larger than Planck mass, $m_{Pl} = (\hbar c/G)^{1/2} \sim 10^{-5}$ grams.

We use the term 'initial system' to refer to the quantum system Q on which a measurement is to be made by a classical apparatus A, and the term 'final system' to refer jointly to Q and A after the initial system has interacted with A. A quantum measurement will be thought of as an increase in the mass (equivalently, number of degrees of freedom) of the system, from the initial value $m_Q \ll m_{Pl}$ to the final value $m_Q + m_A \gg m_{Pl}$. Clearly then, the non-Hermitian part of the Hamiltonian will play a crucial role in the transition from the initial system to the final system.

We assume that A measures an observable $\hat{O}$ of Q, having a complete set of eigenstates $|\psi_n >$. Let the quantum state of the initial system be given as $|\psi >= \Sigma_n a_n|\psi_n >$. The onset of measurement corresponds to mapping the state $|\psi >$ to the entangled state $|\psi >_F$ of the final system as

$$|\psi >\rightarrow |\psi >_F \equiv \sum_n a_n|\psi >_{Fn}= \sum_n a_n|\psi_n > |A_n > \qquad (7.3)$$

where $|A_n >$ is the state the measuring apparatus would result in, had the initial system been in the state $|\psi_n >$. During a quantum measurement the non-Hermitian part of the Hamiltonian dominates over the Hermitian part, and governs the evolution of the state $|\psi >_F$.

7.2.2   The toy model

As a useful and illustrative toy-model, we will consider a special case of the non-linear equation (7.2) with the operator U given by

$$U = i\gamma \sum_n q_n |\psi >_{Fn} < \psi|_{Fn}, \qquad H = 0, \qquad (7.4)$$

where the q_n are random real, positive constants [7]. γ is a constant which will be assumed to be zero before the onset of measurement, and non-zero during the measurement. This is consistent with the assumption, made above, that the non-Hermitean part of the Hamiltonian becomes significant only when the mass of the system becomes comparable to or larger than Planck mass. Thus by system here we mean the 'final system', which jointly refers to the measuring apparatus $\mathcal{A}$ and the initial quantum system $\mathcal{Q}$. The state $|\psi >_{Fn}$ has been defined in (7.3) above. The remaining Hermitean part of the Hamiltonian has been set to zero for simplicity; because here we want to demonstrate how the non-Hermitean part is responsible for the decay of the superpositions initially present in the quantum system $\mathcal{Q}$. Including the Hermitean part will not prevent the breakdown of superposition - it will only make the analysis more complicated.

[We are using Grigorenko's model to illustrate collapse induced by non-linearity, because of its simplicity. However, the essential idea has been put forward much earlier, probably first by Bohm and Bub [12], who suggested a non-linear Schrödinger equation to explain measurement, using hidden variables as random variables. More pertinent to our context is a 1976 paper by Pearle [13] who proposed a non-linear Schrödinger equation with phases of states as random variables, to explain measurement. This insightful and highly readable paper already put forth, in spirit, the non-linear mechanism proposed by us, except that a fundamental origin for the non-linearity was not suggested there. We find it curious that subsequent focus shifted mainly to other dynamical reduction models (not involving the phase as a random variable) and Pearle's original suggestion was largely forgotten. In a sense, our present work revives Pearle's idea, bringing it in as a byproduct of considerations originating in quantum gravity.]

Let us now analyze Eqn. (7.2), with the understanding that the state $|\psi >$ is here $|\psi >_F$. Substituting the form of the Hamiltonian given in (7.4) into the non-linear Schrödinger equation (7.2) gives, after using the expansion for $|\psi >_{Fn}$ given in (7.3),

$$da_n/dt = \gamma a_n (q_n - L) \qquad (7.5)$$

where $L = \Sigma_n q_n |a_n|^2$. Hence we can write

$$\frac{d}{\gamma dt}\left(\ln\frac{a_i}{a_j}\right) = q_i - q_j. \tag{7.6}$$

Since the evolution is norm-preserving, it is easily inferred that the system evolves to the state with the largest value of q_n. In a repeated measurement, different outcomes can be achieved by different sets of values of the random variables q_n. For instance, if the q_n each lie in the range $[0, \infty]$ and have a probability distribution

$$\omega(q_n) = | < \psi(t_0)|\psi_n > |^2 \exp(| < \psi(t_0)|\psi_n > |^2 q_n) \tag{7.7}$$

then the probability that the system evolves to the state $|\psi_n >$ can be shown to be $| < \psi(t_0)|\psi_n > |^2$, as required by experiments [7].

More details and interesting features of this and related models can be found in Grigorenko's paper. For instance, the number of random variables can be less than the number of system states. Furthermore, an attractive candidate for a random variable is the phase of the initial quantum state $\psi(t_0)$. As explained by Grigorenko, as the initial phase varies randomly and uniformly in the range $[0, 2\pi]$, it is possible to have different outcomes in a repeated measurement, depending on the value of the initial phase, and in consistency with the Born probability rule. Grigorenko also discusses how to overcome the problem of superluminality in his model.

The implication of this model is that evolution during a quantum measurement is deterministic. Probabilities enter the picture because of the presence of one or more random variables in a non-linear Schrödinger equation with a non-Hermitean part. Non-linearity breaks superposition. The non-Hermitean part causes decay/growth of different components of the quantum state. The presence of the random variables ensures that in a repeated measurement, different outcomes are realized, and the Born probability rule can be recovered by associating suitable probability distributions with the random variables.

Nevertheless, this is a toy model, and the form of the Hamiltonian is decidedly *ad hoc*. In Section 7.3 we will show that a Hamiltonian with the same features as in this toy model arises from addressing a fundamental incompleteness of quantum mechanics - the presence of an external classical time in the theory. We will hence be able to provide a collapse based explanation of quantum measurement, for reasons having to do with quantum gravity. We also note that the toy model has a limitation that the form of the non-linear operator can only be written down *after* choosing a basis. This limitation will not arise for the non-linear equation derived in Section 7.3.

7.2.3 The Doebner-Goldin equation

In this brief digression we point out an important non-linear Schrödinger equation which belongs to the general class (7.2). This is the Doebner-Goldin equation [14]

$$ i\hbar\frac{\partial\psi}{\partial t} = -\frac{\hbar^2}{2m}\nabla^2\psi + iD\hbar\left(\nabla^2\psi + \frac{|\nabla\psi|^2}{|\psi|^2}\psi\right). \qquad (7.8) $$

Here, D is a real constant. The origin of the D-G equation has been discussed in detail in [14]; its generalizations are discussed by Goldin [15] and also in [16].

With some further assumptions, the D-G equation can be used to explain the collapse of the wave-function [17], in the manner of the toy-model described above. The non-linear Schrödinger equation that we arrive at in the next Section is very similar to the above D-G equation.

7.3 Quantum gravity suggests that quantum mechanics is non-linear

7.3.1 Outline of the approach

We outline below the key steps in the development of our intended non-linear theory:

- There ought to exist a reformulation of quantum mechanics which does not refer to a classical spacetime manifold. This provides a new path to quantum gravity.
- It then follows as a consequence that quantum theory as we know it is a limiting case of a non-linear quantum theory.
- We propose the desired reformulation of quantum mechanics using ideas from noncommutative differential geometry.
- This has implications for the quantum measurement problem: we arrive at a non-linear Schrödinger equation, with the non-linearity becoming significant in the vicinity of the Planck-mass scale.

A detailed discussion of the approach can be found in [16].

7.3.2 Why quantum mechanics without classical spacetime?

The concept of time that appears in quantum mechanics is a classical concept. It is part of a classical spacetime manifold. The overlying metric on this spacetime manifold is produced by classical matter fields. In our present Universe we take the presence of such classical matter fields as given. But the Universe could in principle be in a state in which there are no classical matter fields.

If only quantum matter fields are present in the Universe, then the metric produced by them will undergo quantum fluctuations. If the metric is undergoing quantum fluctuations, then one cannot assign physical significance to the underlying classical spacetime manifold. This is the essence of the Einstein hole argument [18], [19]. Hence it is necessary to have a fundamental reformulation of quantum mechanics which does not refer to a classical spacetime manifold.

[**The Einstein hole argument** : Consider a spacetime manifold $\mathcal{M}$ having matter fields, except in a hole H inside the manifold, which is devoid of matter fields. The only field present in the hole is the gravitational field, described by the metric $g_{\mu\nu}$. Consider an active diffeomorphism ϕ on M, which is by definition identity outside the hole, and on its boundary, but different from identity inside the hole. Clearly, the stress-tensor $T_{\mu\nu}$ remains unchanged under this diffeomorphism, all over the spacetime. However, inside the hole, the metric changes under the diffeomorphism, say from $g_{\mu\nu}(p)$ at the point p to $\phi * g_{\mu\nu}(q)$ at the point q, where q is the point to which the point p is mapped, by the active diffeomorphism. Since $T_{\mu\nu}$ has not changed, it is natural to expect that the physical gravitational field has not changed either. This is achieved by appealing to general covariance, namely that the metrics $g_{\mu\nu}(p)$ and $\phi * g_{\mu\nu}(q)$ describe the same physical gravitational field.

However, Einstein realized that general covariance comes at a price. The fields $g_{\mu\nu}(p)$ and $\phi * g_{\mu\nu}(q)$ can be regarded as physically identical only if the points p and q are also regarded as physically identical! It is thus a consequence of general covariance that points on a spacetime manifold cease to have any physical meaning as separate, distinct points of a spacetime. The only way to restore a physical attribute to points of the spacetime manifold (i.e. consider them as events) is through the presence, on the manifold, of a specific dynamically determined metric tensor field— the metric tensor field serves to provide a label to the points.

In our present context, we hence see that if the metric tensor is undergoing quantum fluctuations, one can no longer regard the underlying classical spacetime manifold as having a physically meaningful status [20].]

As and when the Universe is in a state in which it is dominated by classical matter fields, a classical spacetime manifold, and a classical metric become available. Quantum mechanics could then be equivalently described in two ways: (i) either in the proposed reformulation, which continues to avoid making reference to the time which is now externally available, or, as we always do (ii) from the point of view of an observer in the classical Universe, as standard quantum mechanics wherein evolution is described with respect to an external time. The proposed reformulation should become equivalent to the standard formulation when an external time becomes available.

In such a reformulation there is no classical spacetime; however, we can envisage the concept of a 'quantum spacetime' and a 'quantum gravitational field' which is produced by quantum matter fields. The 'quantum gravitational field', like its classical counterpart, is assumed to act as a source for itself. This makes the quantum theory of gravity a non-linear theory—this feature is central to the thesis of this paper. Such a theory is completely different from the Wheeler-DeWitt equation, wherein the quantum theory of gravity is linear by construction.

A non-linear quantum gravity might appear counterintuitive because we are generally conditioned to building a quantum theory by 'quantizing' a classical theory. These rules of quantization are linear by definition, and further, they assume an external classical time as given. However, when such an external classical time is no longer available, we do not have such rules of quantization ready at hand. We are compelled to adopt a top-down approach, and then we see that it is very natural that 'quantum gravity' 'quantum gravitates', in precisely the same sense in which classical gravity acts as a source for itself—thus leading to a non-linear quantum gravity. What this means, for instance, is that if we were to write an analog of the Wheeler-DeWitt equation for the non-linear case, the Hamiltonian of the theory would depend on the quantum state.

As has been explained in detail in [16], this non-linearity in the quantum gravity theory becomes significant only at the Planck mass/energy scale. At much lower energy scales, the theory is linear, to an excellent approximation. This has an analogy with classical general relativity - the classical theory is non-linear only in the strong-field regime, but linear in the weak field Newtonian regime.

Next, we consider what the equation of motion of a quantum field or a quantum mechanical particle is, in such a 'quantum spacetime'. Again, as shown in detail in [16], if the mass of the particle is comparable to Planck mass, its quantum dynamics is influenced by its own quantum gravitational field, and the equation of motion is non-linear. From the point of view of an external spacetime (as and when the latter becomes available) this equation of motion (in the non-relativistic limit) will appear to be a non-linear Schrödinger equation. When the particle's mass is much smaller than Planck mass, the non-linear equation will reduce to the standard linear Schrödinger equation.

7.3.3 *A reformulation based on noncommutative differential geometry*

We suggest the concept of a noncommuting coordinate system, which 'covers' a noncommutative manifold, wherein commutation relations between coordinates are introduced on physical grounds. Commutation relations amongst momenta must also be introduced.

Our proposal is that basic laws are invariant under general coordinate transformations of noncommuting coordinates.This generalizes the standard concept of general covariance to noncommuting coordinates. This formulation should satisfy two important properties:

- Firstly, in the limit in which the system becomes macroscopic, the noncommutative spacetime should be indistinguishable from ordinary commutative spacetime, and the dynamics should reduce to classical dynamics.
- Secondly, if a dominant part of the system becomes macroscopic and classical, and a sub-dominant part remains quantum (as our Universe is) then seen from the viewpoint of the dominant part, the quantum dynamics of the sub-dominant part should be the same as the standard quantum dynamics known to us.

Our overall proposal for the reformulation of quantum mechanics in the language of noncommutative geometry can be stated as follows:

- Noncommutative special relativity gives the new formulation of relativistic quantum mechanics which does not refer to a classical spacetime. The standard quantum commutation relations of quantum mechanics are deduced from the spacetime and the momentum commutation relations.
- Noncommutative general relativity is quantum gravity.

In the next sub-section we suggest the reformulation of quantum mechanics in terms of noncommuting coordinate systems, and the recovery of standard quantum mechanics from this reformulation. We then show how inclusion of self-gravity leads to a non-linear Schrödinger equation Much work still remains to be done, in terms of making rigorous contact with noncommutative differential geometry, and arriving at the field equations for noncommutative general relativity. However the flow of ideas appears rather natural, and we arrive at an explanation for collapse-induced quantum measurement which can be subjected to an experimental test. We would like to emphasize once again that the ideas in this section were not developed with the *ad hoc* purpose of explaining quantum measurement, but have been concerned with an altogether different aspect of quantum mechanics - the unsatisfactory presence of an external classical time in the theory.

7.3.4 *Quantum Minkowski spacetime*

Consider a system of quantum mechanical particles having a total mass-energy much less than Planck mass m_{Pl}, and assume that no external classical spacetime manifold is available. Since Planck mass scales inversely with the gravitational constant, we are justified here in neglecting the gravitational field, and the resulting quantum spacetime produced by the system will be called a 'quantum Minkowski spacetime'.

To describe the dynamics using noncommutative geometry consider a particle with mass $m \ll m_{Pl}$ in a 2-d noncommutative spacetime with coordinates $(\hat{x}, \hat{t})$. On the quantum Minkowski spacetime we introduce the non-Hermitean flat metric

$$\hat{\eta}_{\mu\nu} = \begin{pmatrix} 1 & 1 \\ -1 & -1 \end{pmatrix} \tag{7.9}$$

and the corresponding noncommutative line-element

$$ds^2 = \hat{\eta}_{\mu\nu} d\hat{x}^\mu d\hat{x}^\nu = d\hat{t}^2 - d\hat{x}^2 + d\hat{t}d\hat{x} - d\hat{x}d\hat{t} \tag{7.10}$$

which is invariant under a generalized Lorentz transformation.

Noncommutative dynamics is constructed by formally defining a velocity $\hat{u}^i = d\hat{x}^i/ds$, which, from (7.10), satisfies the relation

$$1 = \hat{\eta}_{\mu\nu} \frac{d\hat{x}^\mu}{ds} \frac{d\hat{x}^\nu}{ds} = (\hat{u}^t)^2 - (\hat{u}^x)^2 + \hat{u}^t\hat{u}^x - \hat{u}^x\hat{u}^t. \tag{7.11}$$

We define a generalized momentum as $\hat{p}^i = m\hat{u}^i$, which hence satisfies

$$\hat{p}^\mu \hat{p}_\mu = m^2. \tag{7.12}$$

Here, $\hat{p}_\mu = \hat{\eta}_{\mu\nu}\hat{p}^\mu$ is well-defined. Written explicitly, this equation becomes

$$(\hat{p}^t)^2 - (\hat{p}^x)^2 + \hat{p}^t\hat{p}^x - \hat{p}^x\hat{p}^t = m^2. \tag{7.13}$$

Dynamics is constructed by introducing a *complex* action $S(\hat{x}, \hat{t})$ and by defining the momenta introduced above as gradients of this complex action. In analogy with classical mechanics this converts (7.13) into a (noncommutative) Hamilton-Jacobi equation, which describes the dynamics.

When an external classical Universe with a classical manifold (x, t) becomes available (see below), one defines the generalized momentum (p^t, p^x) in terms of the complex action $S(x, t)$ as

$$p^t = -\frac{\partial S}{\partial t}, \qquad p^x = \frac{\partial S}{\partial x} \tag{7.14}$$

and from (7.13) the following fundamental rule for relating noncommutative dynamics to standard quantum dynamics

$$(\hat{p}^t)^2 - (\hat{p}^x)^2 + \hat{p}^t\hat{p}^x - \hat{p}^x\hat{p}^t = (p^t)^2 - (p^x)^2 + i\hbar\frac{\partial p^\mu}{\partial x^\mu}. \tag{7.15}$$

A detailed justification for this key equation has been given in [16].

In terms of the complex action the right hand side of this equation can be written as

$$\left(\frac{\partial S}{\partial t}\right)^2 - \left(\frac{\partial S}{\partial x}\right)^2 - i\hbar\left(\frac{\partial^2 S}{\partial t^2} - \frac{\partial^2 S}{\partial x^2}\right) = m^2 \tag{7.16}$$

and from here, by defining a quantum state ψ in a natural manner: $\psi = e^{iS/\hbar}$, we arrive at the Klein-Gordon equation

$$-\hbar^2\left(\frac{\partial^2}{\partial t^2} - \frac{\partial^2}{\partial x^2}\right)\psi = m^2\psi. \tag{7.17}$$

In this manner we have arrived at standard quantum mechanics, starting from a formulation which did not make reference to a classical time.

The proposed commutation relations on the non-commutative spacetime are

$$[\hat{t}, \hat{x}] = iL_{Pl}^2, \qquad [\hat{p}^t, \hat{p}^x] = iP_{Pl}^2. \tag{7.18}$$

In [16] it has been suggested as to how one could infer the standard commutation relation of quantum mechanics from the relations given above.

We have restricted the discussion here to a single particle in two dimensions. The generalization to the multi-particle case, in four dimensions, is straightforward, and outlined in [16].

7.3.5 *Including self-gravity*

If the mass-energy of the particle is not negligible in comparison to Planck mass its self-gravity must be taken into account. The 'flat' metric (7.9) gets modified to the 'curved' metric

$$\hat{h}_{\mu\nu} = \begin{pmatrix} \hat{g}_{tt} & \hat{\theta} \\ -\hat{\theta} & -\hat{g}_{xx} \end{pmatrix} \tag{7.19}$$

We have made the significant assumption that in addition to the standard symmetric metric $\hat{g}_{\mu\nu}$ the noncommutative 'curved' metric also has an antisymmetric component $\hat{\theta}_{\mu\nu}$. The component $\hat{\theta}_{\mu\nu}$ will play a central role in our explanation of quantum measurement.

With the introduction of the curved metric, Eqns.(7.10), (7.12), (7.15) and (7.16) are respectively replaced by the plausible equations

$$ds^2 = \hat{h}_{\mu\nu}d\hat{x}^{\mu}d\hat{x}^{\nu} = \hat{g}_{tt}d\hat{t}^2 - \hat{g}_{xx}d\hat{x}^2 + \hat{\theta}[d\hat{t}d\hat{x} - d\hat{x}d\hat{t}], \tag{7.20}$$

$$\hat{h}_{\mu\nu}\hat{p}^{\mu}\hat{p}^{\nu} = m^2, \tag{7.21}$$

$$\hat{g}_{tt}(\hat{p}^t)^2 - \hat{g}_{xx}(\hat{p}^x)^2 + \hat{\theta}\left(\hat{p}^t\hat{p}^x - \hat{p}^x\hat{p}^t\right) = m^2, \tag{7.22}$$

$$g_{tt}(p^t)^2 - g_{xx}(p^x)^2 + i\hbar\theta\frac{\partial p^{\mu}}{\partial x^{\mu}} = m^2. \tag{7.23}$$

These replacements have been made very much in the same spirit in which one goes from flat spacetime equations to curved spacetime equations in classical general relativity. Like in general relativity, the metric $\hat{h}_{\mu\nu}$ is assumed to be determined by the mass m via the quantum state $S(\hat{x}, \hat{t})$. The field equations are assumed to be covariant under general coordinate transformations of noncommuting coordinates. If these field equations could be determined, they would constitute the field equations of quantum gravity, in this approach.

In the macroscopic limit $m \gg m_{Pl}$ the antisymmetric component θ is assumed to go to zero; the noncommutative spacetime (7.20) is then indistinguishable from ordinary commutative spacetime, and Eqn. (7.23) reduces to classical dynamics. In the microscopic limit $m \ll m_{Pl}$ we have that θ goes to one, g_{tt} and g_{xx} also go to one, and we recover standard quantum mechanics. Thus we see that when θ is different from zero and one, we get a new mechanics which is neither standard linear quantum mechanics, nor classical mechanics!

There are, however, issues which remain to be resolved. We have assumed θ to be real, which appears a reasonable choice, considering that it represents an additional component of the gravitational field, significant only in the mesoscopic domain. We do not know at the moment the explicit spacetime dependence of θ - we will assume as of now that θ depends only on the mass m, and on the quantum state $S(x, t)$.

7.3.6 A non-linear Schrödinger equation

If we substitute for the momenta in (7.23) in terms of the complex action using (7.14) and then substitute $\psi = e^{iS/\hbar}$ and take the non-relativistic limit, the resulting effective Schrödinger equation is non-linear [16]. It is very similar to the Doebner-Goldin equation discussed before - the latter arises when one classifies physically different quantum systems by considering unitary representations of the group of diffeomorphisms $Diff(R^3)$.

The simplest case is obtained when in (7.23) one approximates the diagonal metric components to unity, giving the non-linear Schrödinger equation

$$i\hbar\frac{\partial \psi}{\partial t} = -\frac{\hbar^2}{2m}\frac{\partial^2 \psi}{\partial x^2} + \frac{\hbar^2}{2m}(1-\theta)\left(\frac{\partial^2 \psi}{\partial x^2} - [(\ln \psi)']^2\psi\right) + V(x)\psi. \quad (7.24)$$

We have generalized by including a potential $V(x)$. This equation bears a striking resemblance to the Doebner-Goldin equation (7.8); the two equations have been compared in detail in [16].

This equation can be rewritten as,

$$i\hbar\frac{\partial \psi}{\partial t} = -\frac{\hbar^2}{2m}\frac{\partial^2 \psi}{\partial x^2} + \frac{\hbar^2}{2m}(1-\theta)\left(\frac{\partial^2[\ln \psi]}{\partial x^2}\right)\psi + V(x)\psi, \quad (7.25)$$

and then more usefully, by expanding the non-linear term into real and imaginary parts, as

$$i\hbar\frac{\partial \psi}{\partial t} = -\frac{\hbar^2}{2m}\frac{\partial^2 \psi}{\partial x^2} + V(x)\psi + \frac{\gamma(m)\hbar^2}{2m}q\frac{\partial^2(\ln R)}{\partial x^2}\psi + i\frac{\gamma(m)\hbar}{2m}q\frac{\partial^2\phi}{\partial x^2}\psi + V(x)\psi$$

$$(7.26)$$

where $\gamma(m)q = (1-\theta)$ and $\psi = Re^{i\phi/\hbar}$. We have made the plausible assumption that $(1-\theta)$ can be written as a product of two positive terms - a part $\gamma(m)$ which does not depend on the state, and a part q which depends on the state, but not on the mass. We see in the second-last term of (7.26) the emergence of the non-Hermitean, non-linear part which is of interest to us. Gravity is responsible for this term because θ is actually a function of m/m_{Pl}, and so is γ.

It may appear that the non-linear equation we have derived is complicated, and does not possess the simplicity of the linear Schrödinger equation. However it is worth recalling that this equation is the non-relativistic limit of the highly symmetric Eqn. (7.21). The relativistic equation, when written in noncommuting coordinates, and in terms of the complex action, has a symmetric and simple form.

The above equation is thus similar to the non-linear Schrödinger equation (7.2) reviewed in the previous section. However, our equation is not norm-preserving! It is norm-preserving if the probability density is defined as $|\psi|^{2/\theta}$, instead of $|\psi|^2$. We need not regard this circumstance as an implausible one, since in this mesoscopic domain (where θ is neither one nor zero) we would not know a priori what the exact definition of norm, in terms of the wave-function, is [16].

One could ask for a reason for the presence of the non-Hermitean term in the Hamiltonian in Eqn. (7.24). The answer is that the presence of such a term is generic; it is only in the small mass, linear, limit that this term is negligible. Further, so long as the evolution is norm-preserving, the presence of such a term cannot be regarded as objectionable.

In passing, we note that in terms of the complex action function S defined earlier as $\psi = e^{iS/\hbar}$ the non-linear Schrödinger equation (7.24) is written as

$$\frac{\partial S}{\partial t} = -\frac{S'^2}{2m} + \frac{i\hbar}{2m}\theta(m)S'' + V(x). \tag{7.27}$$

This equation is to be regarded as the non-relativistic limit of Eqn. (7.23). We easily see here that in the limit $\theta = 0$ classical mechanics is recovered; and that setting $\theta = 1$ gives the linear Schrödinger equation. The intermediate regime, where θ is neither one nor zero, is different from both classical and quantum mechanics. This regime is consistent both with classical and quantum mechanics, but will go undetected in experiments unless one examines properties of mesoscopic systems. Interestingly enough, if θ is different from one, the non-linear equation can explain the collapse of the wave-function, as we will now see.

7.3.7 *Explaining quantum measurement*

Prior to the onset of a quantum measurement, evolution is described by

$$i\hbar\frac{\partial\psi}{\partial t} = -\frac{\hbar^2}{2m}\frac{\partial^2\psi}{\partial x^2} + V(x)\psi \tag{7.28}$$

thus preserving superposition. This is because we have $m \ll m_{Pl}$ and $\theta \to 1$.

We recall that the onset of measurement corresponds to mapping the state $|\psi>$ to the state $|\psi>_F$ of the final system as

$$|\psi>\rightarrow |\psi>_F \equiv \sum_n a_n|\psi>_{Fn}= \sum_n a_n|\psi_n> |A_n> \qquad (7.29)$$

where $|A_n>$ is the state the measuring apparatus would result in, had the initial system been in the state $|\psi_n>$.

At the onset of measurement, evolution is described by the equation

$$i\hbar\frac{\partial\psi_F}{\partial t} = H_F\psi_F + \frac{\gamma(m)\hbar^2}{2m_F}q\frac{\partial^2(\ln R_F)}{\partial x^2}\psi_F + i\frac{\gamma(m)\hbar}{2m_F}q\frac{\partial^2\phi_F}{\partial x^2}\psi_F. \qquad (7.30)$$

H_F is the Hermitean part of the Hamiltonian for the final system (including both quantum system and measuring apparatus). Also, $\gamma(m_F)q = (1-\theta_F)$ and $\psi_F = R_F e^{i\phi_F/\hbar}$. This equation should be compared with the non-linear equation in the toy model.

The states $|\psi>_{Fn}$ cannot evolve as a superposition because the evolution is now non-linear. However, the initial state at the onset of measurement *is* a superposition of the $|\psi_{Fn}>$; it is simply the entangled state $|\psi>_F$ at the onset of measurement. This initial superposition must thus break down during further evolution, according to the law

$$i\hbar\frac{\partial a_n}{\partial t} = i\frac{\gamma(m_F)\hbar}{2m_F}q\frac{\partial^2\phi_F}{\partial x^2}a_n \qquad (7.31)$$

which follows after substituting the expansion (7.29) in (7.30). Here, we ignore the Hermitean part of the Hamiltonian, and focus only on the decay/growth of the quantum state. Note that the q_n's are different for different states. This is because it is natural to assume that the component θ of the gravitational field, which is produced by the mass m, should depend on the quantum state. Also, ϕ_F is the value of the phase of the state $|\psi>_F$ at the onset of measurement.

We thus get, as before

$$\frac{d}{dt}\ln\frac{a_i}{a_j} = \frac{\gamma}{2m_F}(q_i - q_j)\phi_F'' \qquad (7.32)$$

and like for the toy model, only the state with the largest q survives. This is ensured because the a_n satisfy the condition $\Sigma|a_n|^2 = 1$ because of the initial conditions imposed on them. In addition, as noted above, $|\psi|^{2/\theta_F}$, and hence $(\Sigma|a_n|^2)^{1/\theta_F}$ is preserved during evolution. Its important to note the subtlety that there will be a θ_F associated with the state $|\psi>_F$ and different θ_{Fn} associated with each of the states $|\psi>_{Fn}$.

In order to recover the Born rule, the q_n must be random variables. Only further development in theory can determine if this is so, and what their probability distribution is. One could nonetheless, following Grigorenko, assign a probability distribution for the q_n so as to recover the Born rule. Once again, the phase of the initial quantum state $\psi(t_0)$ is an attractive candidate for the desired random variables.

From Eqn. (7.31) we can define an important quantity, the lifetime τ_{sup} of a superposition. It can be read off from this equation to be

$$\tau_{sup} = \frac{2m}{(1 - \theta)\phi_F''}. \tag{7.33}$$

The first inference we can draw is that, since θ is strictly equal to one in standard linear quantum mechanics, a quantum superposition has an infinite lifetime in the linear theory, as one would expect. However, the situation begins to change in an interesting manner as the value of the mass m approaches and exceeds m_{Pl}. Since we know that in this limit θ approaches zero, we can neglect θ, and the superposition lifetime will then essentially be given by

$$\tau_{sup} \approx \frac{m}{\phi_F''} \sim \frac{mL^2}{\phi} \tag{7.34}$$

where L is the linear dimension of the system, and ϕ_F is the phase of the state $|\psi>_F$ at the onset of measurement. For a macroscopic system we can get a numerical estimate of the life-time by noting that we are close to the classical limit, where the phase coincides with the classical action in the Hamilton-Jacobi equation. To leading order, the magnitude of the classical action is given by $S_{cl} = mc^2t$, where t is the time over which we observe the classical trajectory; approximately, this could be taken to be the value of the phase ϕ, and τ_{sup} is then roughly given by

$$\tau_{sup} \sim \frac{1}{t}\left(\frac{L}{c}\right)^2. \tag{7.35}$$

For a measuring apparatus, if we take the linear dimension to be say 1 cm, and the time of observation to be say 10^{-3} seconds, we get the superposition lifetime to be 10^{-18} seconds, which is an encouragingly small number. This could possibly explain why the wave-function collapses so rapidly during a measurement.

We can get a very rough estimate of τ_{sup} for a mesoscopic system using (7.35), and continuing to ignore θ. Let us take $L \sim 10^{-3}$ cm and correspondingly $m \sim 10^{-9}$ gm. Such a composite object has approximately 10^{15}

particles, and we could take $\phi \sim N\hbar$ with $N \sim 10^{15}$. This gives $\tau_{sup} \sim 10^{-3}$ seconds.

Thus, in making a transition from a microscopic system which obeys linear quantum mechanics, to a macroscopic system such as a measuring apparatus, we find that the lifetime of a superposition changes from an astronomically large value to an immeasurably small value. We could thus be certain that there must exist intermediate, mesoscopic systems for which the lifetime of a superposition is an easily measurable number, say one millisecond. Unfortunately our present understanding of the approach described here is not good enough to say at what value of the mass m this will happen. What we can be certain about is that if the ideas described here are on the right track, then as experimentalists check for quantum superposition in larger and larger quantum systems, they will discover systems for which the lifetime of a superposition will become small enough to be measurable in the laboratory in principle. In practice, it remains to be seen whether or not decoherence will permit this lifetime to be measured.

7.3.8 *Ideas for an experimental test of the model*

The possibilities for an experimental test of this non-linear model, and of its implications for quantum measurement, can be divided into three classes:

(i) Looking for breakdown of quantum superposition: This suggestion is in line with the kind of experiments already in progress - constructing larger and larger composite quantum objects (i.e. Carbon-60 and beyond) and checking whether or not linear superposition holds for such objects. Our prediction is that by the time the number of particles in the composite object reaches to about 10^{15} particles, the lifetime of the superposition will become small enough to be observable in the laboratory, and one will actually observe the decay of superposition in such a system. The greatest obstacle to the detection of such an effect, even if it is there, will come from the phenomenon of decoherence. In order to ascertain whether or not there is a breakdown of non-linearity due to superposition, one will have to first ensure that the object is sufficiently well-isolated from its environment, so that decoherence can be avoided.

(ii) Difference between the predictions of the linear theory and the non-linear theory: If one were to calculate expectation values of observables, using the non-linear equation (7.24), the result will be different from what one will get from the linear theory, because now θ is non-zero. This aspect has been discussed in some detail in [16]. Care has to be taken to construct gauge-invariant observables for the non-linear theory. It was shown in [16] that the ratio $\hbar/m$ is one such gauge-invariant observable - the non-linear model predicts that the effective Planck's constant in the theory is $\hbar\theta(m)/m$. Thus, according to the non-linear model, an experiment to measure the value of $\hbar/m$ for a mesoscopic system will give results consistent with theory only if one assumes Planck's constant to have an effective value $\hbar\theta(m)$.

(iii) Looking for a correlation between the absolute value of the initial phase, and the outcome of a quantum measurement : If the random variable responsible for the outcome of a quantum measurement in the non-linear Schrödinger equation (7.30) is indeed the absolute phase of the initial state $\psi(t_0)$, then the correlation between this phase and the outcome of a measurement should in principle be detectable by experiment. Conventional wisdom is that absolute phase cannot be measured and that only differences in phase are measurable. However, if the absolute value of the phase is going to decide the outcome of a measurement, it does not seem unnatural to expect that there might be a way to determine the value of this initial phase. In other words, while absolute phase is not observable in the linear theory, it probably does become observable in the non-linear theory, and one should explore possible ways to measure it in the non-linear theory, i.e. for mesoscopic systems.

7.4   Other models for collapse of the wave-function

Various researchers have proposed modifications of the Schrödinger equation, with the purpose of explaining quantum measurement as a dynamically induced collapse. Most of these models also estimate the lifetime of a quantum superposition - the lifetime being large for microscopic systems, and small for macroscopic ones. While some models do not involve gravity, it is remarkable that many of the models assert that gravity is responsible for collapse. Below, we review some of the models very briefly, without really attempting to critically compare these models with each other, or with our own approach. (A comparison across various models of collapse will be reported elsewhere. The literature on the subject is large, and we confine ourselves to giving pointers to the literature. A nice, though somewhat older, review of collapse models has been given by Pearle [21]).

7.4.1 *Models that do not involve gravity*

Dynamical reduction models based on non-linear stochasticity:
The first suggestion that collapse could be explained as a dynamical reduction process using stochastic differential equations came from Pearle [22]. The Schrödinger equation was to be augmented by a stochastic term which could induce collapse. Significant development in this direction came from the work of Ghirardi, Rimini and Weber (GRW) [23]. This program, and its progress, has been reviewed in [24] and in [25]. There were two guiding principles for this dynamical reduction model (known as QMSL: Quantum Mechanics with Spontaneous Localization) [24] :

" 1. The preferred 'basis' - the basis on which reductions take place - must be chosen in such a way as to guarantee a definite position in space to macroscopic objects.

2. The modified dynamics must have little impact on microscopic objects, but at the same time must reduce the superposition of different macroscopic states of macro-systems. There must then be an amplification mechanism when moving from the micro to the macro level. "

The reduction is achieved by making the following set of assumptions:

" 1. Each particle of a system of n distinguishable particles experiences, with a mean rate λ_i, a sudden spontaneous localization process.

2. In the time interval between two successive spontaneous processes the system evolves according to the usual Schrödinger equation. "

In their model, GRW introduced two new fundamental constants of nature, assumed to have definite numerical values, so as to reproduce observed features of the microscopic and macroscopic world. The first constant, $\lambda^{-1} \sim 10^{16}$ seconds, alluded to above, determines the rate of spontaneous localization (collapse) for a single particle. For a composite object of n particles, the collapse rate is $(\lambda n)^{-1}$ seconds. The second fundamental constant is a length scale $a \sim 10^{-5}$ cm which is related to the concept that a widely spaced wave-function collapses to a length scale of about a during the localization.

A gravity based implementation of the QMSL model has been studied by Diosi [29] and generalized in [30].

The QMSL model has the limitation that it does not preserve symmetry of the wave-function under particle exchange, and has been improved into what is known as the CSL (Continuous Spontaneous Localization) model [26], [28]. In CSL a randomly fluctuating classical field couples with the particle number density operator of a quantum system to produce collapse towards its spatially localized eigenstates. The narrowing of the wavefunction amounts to an increase in the particle's energy, and actually amounts to a violation of energy conservation. This intriguing aspect of the collapse model has been discussed in [31] and an interesting suggestion for preserving energy conservation has been made therein.

An outstanding open question with regard to the dynamical reduction models is the origin of the random noise, or the randomly fluctuating classical scalar field, which induces collapse. We see herein the possibility of a connection, worth exploring further, with our proposal that the non-linear Schrödinger equation has its origin in quantum gravity. Could the randomly fluctuating classical field of CSL be related to the random parameter $\theta(m)$ we have in our non-linear equation?

Another collapse model is due to Adler (see [27] and related references therein), where an energy-driven stochastic Schrödinger equation is a phenomenological model for state-vector reduction. Collapse in this model is energy conserving and reproduces the Born probability rule. It is interesting to note that in this model, the terms that influence wave-vector reduction are directly related to mass-energy, and hence once again a connection with gravity is suggested.

7.4.2 *Models that involve gravity*

The gravitational field produced by a classical object obeys the laws of general relativity, or in the limiting case, those of Newtonian gravity. Since the position of the classical object is subject to tiny quantum uncertainties, the gravitational field and the curvature tensor produced by it are also subject to quantum fluctuations. Karolyhazy [32], [33] built an interesting and plausible model to explain how quantum superposition could be destroyed in macroscopic objects, as a result of these quantum fluctuations in the gravitational field of the object. The possible connection of such a model with the CSL approach has been discussed in [21]. Other models discussing the possible role of gravity in collapse are presented in [34], [35].

In spirit and concept, our work here comes closest to Penrose's idea that gravity is responsible for wave-vector reduction [36]. Penrose argues convincingly that the principle of general covariance and the principle of linear superposition in quantum mechanics are in direct conflict with each other. Our starting point has been essentially the same - we argued that because of general covariance, one cannot have a classical spacetime manifold coexisting with a Universe which has only quantum matter fields. This lead us to conclude that if gravity cannot be neglected, quantum theory must be non-linear - which is essentially what Penrose has argued: general covariance and linear superposition are incompatible with each other.

Penrose develops an estimate for the lifetime of the quantum superposition of two different positions of a macroscopic object, and demonstrates it to be of the order $\hbar/\Delta E$, where ΔE is the gravitational self-energy of the difference between the mass distributions of each of the two locations of the macroscopic object. An experimental test of Penrose's idea has been proposed in [37].

7.5 Discussion

In putting forth the two possible explanations of quantum measurement, we have assumed that the wave-function describes an individual quantum system, and not a statistical ensemble of quantum systems. Also, we have ignored making any mention of the Copenhagen interpretation, which essentially states that upon measurement, the wave-function collapses into one of the eigenstates, but the interpretation does not suggest any mechanism for the collapse.

We believe that our case for the necessity of a reformulation of quantum mechanics is robust. Equally robust is the inference that the standard linear quantum theory is a limiting case of a non-linear quantum theory. However, only partial justification can be given for the use of noncommutative differential geometry for constructing such a reformulation. Having accepted to work in the framework of noncommutative geometry, we regard it as highly attractive that linear quantum theory, and its nonlinear generalization, are respectively the equations of motion in a noncommutative special relativity, and in its generalization to a noncommutative general relativity. Various issues here remain to be understood much better. These include: (i) the physical meaning associated with the noncommutative metric (7.9), (ii) the nature of commutation relations to be imposed on the 'curved' noncommutative metric $\hat{h}_{\mu\nu}$, (iii) the full development of the concept of general covariance in noncommutative geometry, and the related generalization of the concept of curvature (for a review of the current status of this aspect see for instance [38, 39]), (iv) the field equations, analogous to those in general relativity, for this metric, (v) the justification for retaining $\theta(m)$ in the description of the nonlinear Schrödinger equation (7.24), while ignoring the diagonal components $g_{\mu\nu}$.

.

In spite of these important issues which are yet to be addressed and resolved, we regard it as a natural consequence of the required reformulation that there is a new mechanics in the intermediate, mesoscopic domain; and that classical mechanics, as well as linear quantum mechanics, are its limiting cases. The limits are obtained by letting $\theta \to 0$ and $\theta \to 1$ respectively.

We re-emphasize that there is no reason to believe a priori that quantum mechanics continues to hold unchanged in the mesoscopic domain. And if there indeed are reasons to expect a departure from standard quantum theory in this domain (and these reasons are independent of the issue of quantum measurement) then experiments must be carried out to test the laboratory predictions of these new ideas.

There are very stringent experimental bounds on the presence of non-linear terms in quantum mechanics in the atomic domain [40, 41, 42]. However, these bounds do not extend to the mesoscopic domain—thus for instance there are no bounds on non-linear quantum mechanics when such a theory is applied to an object containing say 10^{15} particles.

Superluminality: It has also been pointed out that the presence of a non-linearity in quantum mechanics can result in the possibility of superluminal communication [43, 44]. In our approach, the non-linearity is a relic of a more fundamental description of the theory in terms of a noncommutative spacetime. We do not have at present a good understanding of the 'light-cone structure' in a noncommutative spacetime, and it is difficult to assess the possibility or otherwise of superluminal communication in a noncommutative geometry. More importantly, the non-linearity we predict becomes significant only in the mesoscopic domain, and we are suggesting that mechanical laws here are different from both the classical and the quantum case. Thus in this domain the issue of superluminality needs to be addressed afresh. Once again, it needs to be stressed that the effects of non-linearity could be severely masked by decoherence resulting from interaction with the environment, making the mesoscopic mechanics effectively indistinguishable from classical or quantum mechanics.

With regard to superluminality, it has also been pointed out by Doebner and Goldin [45] that by virtue of non-linear gauge transformations, many non-linear Schrödinger equations are physically equivalent to linear equations. So one could not possibly deduce superluminal communication for such non-linear equations.

The preferred basis problem: We would like to suggest, following GRW, that a preferred basis is one in which positions of macroscopic objects are localized. This is consistent with the oft expressed view that *"all quantum mechanical measurements consist of or are obtained from positional measurements made at various times [45]"*. As stated by Feynman and Hibbs [46], p. 96:

"Indeed all measurements of quantum mechanical systems could be made to reduce eventually to position and time measurements. Because of this possibility a theory formulated in terms of position measurements is complete enough in principle to describe all phenomena".

Acknowledgements

For useful discussions and conversations during various stages of this work, it is a pleasure to thank Joy Christian, Patrick Das Gupta, Atri Deshamukhya, Avinash Dhar, Hanz-Dietrich Doebner, Thomas Filk, Gerald Goldin, T. R. Govindarajan, Kumar S. Gupta, Sashideep Gutti, Friedrich Hehl, Eric Joos, F. Karolyhazy, Romesh Kaul, Claus Kiefer, Gautam Mandal, Shiraz Minwalla, Ayan Mukhopadhyay, T. Padmanabhan, Aseem Paranjape, Rainer Plaga, L. Sriramkumar, Rakesh Tibrewala, A. V. Toporensky, Sandip Trivedi, C. S. Unnikrishnan, Cenalo Vaz, Albrecht von Müller and Spenta Wadia.

References

[1] Hugh Everett, III *"Relative State" Formulation of Quantum Mechanics*, Rev. Mod. Phys. 29, 454 (1957).

[2] David Deutsch and David Wallace, *Probability in the Everett Interpretation: State of Play*, http://pirsa.org/07090062.

[3] R. V. Buniy. S. D. H. Hsu and A. Zee, *Discreteness and the origin of probability in quantum mechanics*, Phys. Lett. B640, 219 (2006) [arXiv:hep-th/0606062]..

[4] Max Tegmark, *Many Lives in Many Worlds*, Nature 448, 23 (2007).

[5] M. Brune, E. Hagley, J. Dreyer, X. Maitre, A. Maali, C. Wunderlich, J. M. Raimonde and S. Haroche, *Observing the progressive decoherence of the "meter" in a quantum measurement*, Phys. Rev. Lett. 77, 4887 (1996).

[6] E. Joos, H. D. Zeh, C. Kiefer, D. Giulini, J. Kupsch and I.-O. Stamatescu, *Decoherence and the Appearance of a Classical World in Quantum Theory*, (Sprnger, New York) 2nd Edn.

[7] A. N. Grigorenko, *Measurement description by means of a non-linear Schrödinger equation*, J. Phys.A: Math. Gen. 28, 1459 (1995).

[8] D. Bohm, *A suggested interpretation of the quantum theory in terms of "hidden" variables. I* Phys. Rev. 85, 166 (1952).

[9] M. Gell-Mann and J. B. Hartle, *Classical equations for quantum systems*, Phys. Rev. D47, 3345 (1993).

[10] R. Omnes, *Consistent interpretations of quantum mechanics*, Rev. Mod. Phys. 64, 339 (1992).

[11] N. Gisin, *A simple nonlinear dissipative quantum evolution equation*, J. Phys. A: Math. Gen. 14, 2259 (1981).

[12] D. Bohm and J. Bub, *A proposed solution of the measurement problem in quantum mechanics by a hidden variable theory*, Rev. Mod. Phys. 38, 453 (1966).

[13] P. Pearle, *Reduction of the state-vector by a non-linear Schödinger equation*, Phys. Rev. D13, 857 (1976).

[14] H.-D. Doebner and Gerald A. Goldin, *On a general non-linear Schrödinger equation admitting diffusion currents*, Physics Letters A 162, 397 (1992).

[15] Gerald A. Goldin, *Perspectives on nonlinearity in quantum mechanics*, [arXiv:quant-ph/0002013 (2000)].

[16] T. P. Singh, *Quantum mechanics without spacetime - a case for non-commutative geometry*, in *Quantum theory and symmetries*, Ed. V. K. Dobrev (Heron Press, 2006) [arXiv:gr-qc/0510042].

[17] T. P. Singh (2007), *The inevitable non-linearity of quantum gravity falsifies the many-worlds interpretation of quantum mechanics*, arXiv: 0705.2357 [gr-qc]

[18] A. Einstein (1914), *Die formale grundlage der allgemeinen relativitatstheorie* Koniglich Preussische Akademie der Wissenschaften (Berlin) Sitzungsberichte : 1030-1085.

[19] J. Christian, *Why the quantum must yield to gravity?*, in 'Physics meets Philosophy at the Planck Scale', (Eds. C. Callender and N. Huggett, 1998, Cambridge University Press) [arXiv: gr-qc/9810078].

[20] S. Carlip, *Quantum Gravity: A Progress Report*, Rept. Prog. Phys. 64, 885 (2001) [arXiv:gr-qc/0108040].

[21] P. Pearle, *Collapse Models*, in 'Open Systems and Measurement in Relativistic Quantum Theory', F. Petruccione and H.P. Breuer eds. (Springer Verlag, 1999). [arXiv: quant-ph/9901077].

[22] P. Pearle, *Toward explaining why events occur*, Intl. Jour. Theor. Phys. 18, 489 (1979).

[23] G. C. Ghirardi, A. Rimini and T. Weber, *Unified dynamics for microscopic and macroscopic systems*, Phys. Rev. D34, 470 (1986).

[24] A. Bassi and G. C. Ghirardi, *Dynamical reduction models*, Phys. Rept. 379, 257 (2003) [quant-ph/0302164].

[25] A. Bassi, *Dynamical reduction models : present status and future developments*, J. Phys. Conf. Ser. 67, 012013 (2007) [arXiv: quant-ph/0701014].

[26] P. Pearle, *Combining stochastic dynamical state-vector reduction with spontaneous localization*, Phys. Rev. A39, 2277 (1989).

[27] S. Adler, *Weisskopf-Wigner decay theory for the energy driven stochastic Schrödinger equation*, Phys. Rev. D67, 025007 (2003).

[28] G. C. Ghirardi, A. Pearle and P. Rimini, *Markov processes in Hilbert space and continuous spontaneous localization of systems of identical particles*, Phys. Rev. A42, 78 (1990).

[29] L. Diosi, *Models for universal reduction of macroscopic quantum fluctuations*, Phys. Rev. A40, 1165 (1989).

[30] G. C. Ghirardi, R. Grassi and A. Rimini, *Continuous-spontaneous-reduction model involving gravity*, Phys. Rev. A42, 1057 (1990).

[31] P. Pearle, *Wavefunction collapse and conservation laws*, Found. Phys. 30, 1145 (2000) [arXiv:quant-ph/0004067].

[32] F. Karolyhazy, *Gravitation and quantum mechanics of macroscopic objects*, Nuovo Cimento 42A, 390 (1966).

[33] F. Karolyhazy, A. Frenkel and B. Lukacs, *On the possible role of gravity in the reduction of the wave function* in 'Quantum concepts in space and time', Eds. R. Penrose and C. J. Isham (Oxford University Press, 1986).

[34] I. C. Percival, *Quantum spacetime fluctuations and primary state difusion*, Proc. Roy. Soc. A451, 503 (1995).

[35] D. I. Fivel, *Dynamical reduction theory of Einstein-Podolsky-Rosen correlations and a possible origin of CP-violation*, Phys. Rev. A56, 146 (1997).

[36] R. Penrose, *On gravity's role in quantum state reduction*, Gen. Rel. Grav. 28, 581 (1996).

[37] W. Marshall, C. Simon, R. Penrose and D. Bouwmeester, *Towards quantum superpositions of a mirror*, Phys. Rev. Lett. 91, 130401 (2003).

[38] A. Connes, *Noncommutative geometry : Year 2000*, math/0011193.

[39] J. Madore, *An introduction to noncommutative differential geometry and its physical applications*, Cambridge University Pres (1999).

[40] S. Weinberg, *Precision tests for quantum mechanics*, Phys. Rev. Lett. 62, 485 (1989).

[41] S. Weinberg, *Testing quantum mechanics*, Ann. Phys. 194, 336 (1989).

[42] J. J. Bollinger, D. J. Heinzin, W. M. Itano, S. L. Gilbert and D. J. Wineland, *Test of the linearity of quantum mechanics by rf spectroscopy of the $^9Be^+$ ground state*, Phys. Rev. Lett. 63, 1031 (1989).

[43] N. Gisin, *Stochastic quantum dynamics and relativity*, Helv. Phys. Acta 62, 363 (1989).

[44] J. Polchinski, *Weinberg's non-linear quantum mechanics and the Einstein-Podolsky-Rosen paradox*, Phys. Rev. Lett. 66, 397 (1991).

[45] H.-D. Doebner and Gerald A. Goldin, *Introducing nonlinear gauge transformations in a family of non-linear Schrödinger equations*, Phys. Rev. A 54, 3764 (1996).

[46] R. P. Feynman and A. R. Hibbs, *Quantum Mechanics and Path Integrals*, (McGraw Hill, New York, 1965).

Chapter 8

On the generation and evolution of perturbations during inflation and reheating

L. Sriramkumar

Harish-Chandra Research Institute,
Chhatnag Road, Jhunsi, Allahabad 211 019, India.
E-mail: *sriram@hri.res.in*

Abstract: In this article, I present an introductory overview of inflation and reheating as well as the generation and evolution of perturbations during these epochs. After an outline of inflation and essential, linear, cosmological perturbation theory, I describe the generation of perturbations during inflation. I then sketch as to how reheating is typically achieved with the help of a coarse grained decay rate, and discuss certain effects of post-inflationary dynamics on the evolution of the perturbations.

8.1 Inflation and reheating

Inflation—which refers to a period of accelerated expansion during the early stages of the radiation dominated epoch—is currently considered the most promising paradigm to describe the origin of perturbations in the early universe (see any of the following texts [1] or one of the following reviews [2]). Originally, inflation was proposed to explain the extent of homogeneity and isotropy of the universe [3]. However, very soon, it was realized that the inflationary scenario can also provide an elegant, causal mechanism to generate the initial inhomogeneities [4]. Typically, scalar fields are used to drive inflation, and it is the quantum fluctuations associated with these 'inflatons'[1] that act as the seeds for the primordial inhomogeneities. The inflationary epoch converts the tiny quantum fluctuations present at the be-

---

[1] It is common to refer to the scalar fields that are used to drive inflation as the inflaton.

ginning of the epoch into classical perturbations, which leave their imprints as anisotropies in the Cosmic Microwave Background Radiation (CMBR). These anisotropies in turn evolve due to gravitational instability into the large scale structures that we observe around us today. Even the simplest of scalar field models allows for a sufficiently long period inflation, as is required to account for the extent of homogeneity and isotropy of the universe. Further, in most of the models, the inflaton 'rolls slowly' down the potential, a behavior which leads to a featureless, nearly scale invariant, scalar perturbation spectrum [1, 2]. Such a scalar power spectrum, along with the assumption of a spatially flat, concordant ΛCDM background cosmological model, provides an exceptionally good fit to the recent observations of the anisotropies in the CMBR by missions such as, say, the Wilkinson Microwave Anisotropy Probe (WMAP) [5].

The inflationary phase is expected to occur at energies much higher than, say, the scales associated with the epoch of big bang nucleosynthesis. In the conventional picture, it is assumed that the scalar field that drives inflation does not interact with radiation during the inflationary epoch[2]. As a result, due to the rapid expansion, during inflation, the radiation cools down tremendously, with its temperature falling by the same large factor that the universe expands during this regime. Therefore, at the end of inflation, the universe needs to be rapidly reheated to the original temperature in order to ensure that the key features of the hot big bang model, such as the synthesis of the light elements, remain unaffected. So, according the prevalent view, inflation is expected to be followed by a short period of reheating, an epoch during which the universe attains virtually the same temperature as it had at the beginning of inflation. This is achieved by the transfer of the energy from the inflaton to radiation, and the transfer is supposed to take place due to the coupling between the inflaton and the fields describing the particles that constitute the standard model. However, since an inflationary model that is well motivated from the perspective of high energy physics still eludes us, the coupling remains ill understood, and its form is usually chosen by hand. In fact, often a much simpler approach is adopted wherein the transfer from the inflaton to the radiation is accomplished by introducing a coarse grained decay rate (see Refs. [7]; however, in this context, also see Refs. [8]). As we shall see, the coupling between the inflaton and radiation, even as it allows the universe to reheat,

---

[2]It should be added that there exist other scenarios such as warm inflation wherein this is not true, and the inflaton is considered to be coupled to radiation even during inflation (see, for instance, Refs. [6]).

opens up interesting possibilities for the post-inflationary evolution of the perturbations.

This article aims to provide an overview of inflation and reheating, as well as the origin and the evolution of the perturbations during these two epochs. The plan of the article is as follows. In the following section, I shall outline the way in which inflation provides a large causally connected patch for the generation of the inhomogeneities, and also describe how scalar fields can source inflation. In Sec. 8.3, I shall set up gauge invariant, *linear*[3], cosmological perturbation theory and derive the equations governing the evolution of the scalar, the vector and the tensor perturbations. I shall also discuss as to how the scalar perturbations evolve in the super-Hubble limit. In Sec. 8.4, I shall consider the generation of perturbations during the inflationary regime, and illustrate that slow roll inflation leads to a nearly scale invariant scalar and tensor perturbation spectra. In Sec. 8.5, I shall outline as to how reheating can be achieved through a coarse grained decay rate. In Sec. 8.6, I shall arrive at the equations describing the evolution of the large scale perturbations during reheating, and show as to how perturbative reheating does not affect the amplitude of the large scale curvature perturbations. In Sec. 8.7, I describe the modulated reheating and the curvaton scenarios which exemplify the non-trivial effects that certain post-inflationary dynamics can have on the perturbations. I conclude with Sec. 8.8, which contains a brief summary and discussion.

Before I proceed, a few words on my conventions and notations are in order. I shall work in $(3 + 1)$-dimensions, and adopt the metric signature of $(+, -, -, -)$. While the Greek and Latin indices shall denote the spacetime and the spatial coordinates, respectively, I shall reserve the sub-script k to denote the wavenumber of the perturbations. I shall set $\hbar = c = 1$, but shall display G explicitly, and define the Planck mass to be $M_{\mathrm{p}} = (8 \pi G)^{-1/2}$. I shall express the various quantities in terms of either the cosmic time t or the conformal time η, as is convenient. An overdot and an overprime shall denote differentiation with respect to the cosmic and the conformal time coordinates of the Friedmann metric that describes the expanding universe. It is useful to note here that, for any given function, say, f, $\dot{f} = (f'/a)$ and $\ddot{f} = [(f''/a^2) - (f' a'/a^3)]$, where a is the scale factor associated with the Friedmann metric. Lastly, since observations indicate that the universe has a rather small curvature, as it is usually done in the

---

[3]I should stress here that I shall restrict my attention in this article to perturbation theory at the first order. I shall briefly comment about perturbations at higher orders and its implications for possible deviations from Gaussianity in the concluding section.

context of inflation, I shall work with the spatially flat Friedmann model.

8.2   Inflating the universe

The exceedingly isotropic and perfectly thermal CMBR is a relic radiation that streams to us virtually unimpeded from the epoch of decoupling, when the photons ceased to interact with matter. Though quite isotropic, the CMBR possesses small anisotropies (of about one part in 10^5), and it is the pattern of these fluctuations that contains clues to the physics of the early universe. In the hot big bang model, at the epoch of decoupling, the linear dimension of the forward light cone evolved from the big bang is about 70 times smaller than the backward light cone viewed from today [1, 2]. In other words, the CMBR photons arriving at us from regions of the sky that are sufficiently widely separated could not have interacted by the time of decoupling. Despite this fact, we find that the temperature of the CMBR photons reaching us even from the antipodal directions hardly differ, a drawback of the hot big bang model that is often referred to as the horizon problem. The success of the inflationary paradigm rests on its ability to not only provide a resolution to the horizon problem and thereby explain the extent of isotropy of the CMBR, but also offer a mechanism to generate the anisotropies superimposed upon the smooth background. In the remainder of this section, I shall outline as to how inflation can help us overcome the horizon problem, and also discuss as to how scalar fields can source inflation.

8.2.1   *Drawing the modes back inside the Hubble radius*

There is also a different way of looking at the horizon problem. Consider a spatially flat, smooth, Friedmann universe described by line element

$$ds^2 = dt^2 - a^2(t)\, dx^2 = a^2(\eta)\left(d\eta^2 - dx^2\right), \qquad (8.1)$$

where t is the cosmic time, $a(t)$ is the scale factor, and $\eta = \int [dt/a(t)]$ denotes the conformal time coordinate. Note that the physical wavelengths associated with the perturbations, say, λ_P, always grow as the scale factor, i.e. $\lambda_P \propto a$. In contrast, in a power law expansion of the form $a(t) \propto t^q$, the Hubble radius[4], viz. $d_H = H^{-1} = (\dot{a}/a)^{-1}$, goes as $a^{1/q}$, so that we

---

[4]I consider the Hubble radius rather than the horizon size since, in any power law expansion with $q < 1$, the Hubble radius is equivalent to the horizon up to a finite multiplicative constant. Moreover, as it a local quantity, the Hubble radius often turns out to be more convenient to handle than the horizon.

have $(\lambda_{\mathrm{P}}/d_{\mathrm{H}}) \propto a^{[(q-1)/q]}$. This implies that, when $q < 1$—a condition which applies to both the radiation and the matter dominated epochs—the Hubble radius *shrinks faster* than the physical wavelength *as we go back in time* [1, 2]. In other words, length scales of cosmological interest today (say, $1 \lesssim \lambda_0 \lesssim 10^4$ Mpc), enter the Hubble radius either during the radiation or the matter dominated epochs, and are outside the Hubble radius at earlier times.

If a causal mechanism is to be responsible for the origin of the inhomogeneities, then, clearly, the scales of cosmological interest should be inside the Hubble scales (i.e. $\lambda_{\mathrm{P}} < d_{\mathrm{H}}$) in the very early stages of the universe. This will be possible provided we have an epoch in the early universe during which λ_{P} *decreases faster* than the Hubble radius *as we go back in time*, i.e. if we have [1, 2]

$$-\frac{\mathrm{d}}{\mathrm{d}t}\left(\frac{\lambda_{\mathrm{P}}}{d_{\mathrm{H}}}\right) < 0, \tag{8.2}$$

which then leads to the condition that

$$\ddot{a} > 0. \tag{8.3}$$

In other words, the universe needs to undergo a phase of inflationary (i.e. accelerated) expansion during the early stages of the radiation dominated epoch if a physical mechanism is to account for the generation of the primordial fluctuations.

Let a_{i} and a_{f} denote the scale factors at the beginning and end of inflation. It can be shown that, during inflation [1, 2], the universe has to expand by a factor of about $A = (a_{\mathrm{f}}/a_{\mathrm{i}}) \simeq 10^{26}$ in order to ensure that the forward light cone at decoupling is at least as large as the backward light cone. Given the scale factor, the amount of expansion that has occurred from an initial time t_{i} to a time t is usually expressed in terms of the number of e-folds defined as follows:

$$N = \int_{t_{\mathrm{i}}}^{t} \mathrm{d}t\, H = \ln\left(\frac{a(t)}{a_{\mathrm{i}}}\right), \tag{8.4}$$

where, as mentioned above, $H = (\dot{a}/a)$ is the Hubble parameter. Since $\ln A \simeq \ln 10^{26} \simeq 60$, it is often said that one requires about 60 e-folds of inflation to overcome the horizon problem[5].

---

[5]In fact, 60 e-folds proves to be roughly an observational upper bound which ensures that the largest scale today is inside the Hubble radius during the inflationary epoch [9]. Actually, the number of e-folds needed to resolve the horizon problem depends on the energy scale at which inflation takes place. For instance, if inflation is assumed to occur at a rather low energy scale of, say, 10^2 GeV, then, even 50 e-folds will suffice to surmount the horizon problem.

8.2.2 *Propelling accelerated expansion with scalar fields*

If ρ and p denote the energy density and pressure of the smooth component of the matter field that is driving the expansion, then the Einstein's equations corresponding to the line element (8.1) result in the following two Friedmann equations for the scale factor $a(t)$:

$$H^2 = \left(\frac{8\pi G}{3}\right)\rho \quad \text{and} \quad \left(\frac{\ddot{a}}{a}\right) = -\left(\frac{4\pi G}{3}\right)(\rho + 3p). \qquad (8.5)$$

It is clear from the second Friedmann equation that, for $\ddot{a}$ to be positive, we require that $(\rho + 3p) < 0$. Neither ordinary matter (corresponding to $\rho > 0$ and $p = 0$) nor radiation [which corresponds to $\rho > 0$ and $p = (\rho/3)$] satisfy this condition. In such a situation, we need to identify another form of matter to drive inflation.

Scalar fields, which are often encountered in various models of high energy physics, can easily help us achieve the necessary condition, thereby leading to inflation. Consider a canonical scalar field, say, ϕ, that is described by the standard action and a potential $V(\phi)$. The stress-energy tensor associated with such a scalar field is given by

$$T^{\mu\nu}_{(\phi)} = \partial^\mu \phi \, \partial^\nu \phi - \left[\left(\frac{1}{2}\right)(\partial_\lambda \phi \, \partial^\lambda \phi) - V(\phi)\right] g^{\mu\nu}. \qquad (8.6)$$

The symmetries of the Friedmann background—viz. homogeneity and isotropy—imply that the scalar field will depend only on time and, hence, the resulting stress-energy tensor will be diagonal. Therefore, the energy density ρ_ϕ and the pressure p_ϕ associated with the scalar field simplify to

$$T^0_{0\,(\phi)} = \rho_\phi = \left[\left(\frac{\dot{\phi}^2}{2}\right) + V(\phi)\right], \quad T^i_{j\,(\phi)} = -p_\phi\, \delta^i_j = -\left[\left(\frac{\dot{\phi}^2}{2}\right) - V(\phi)\right]\delta^i_j. \qquad (8.7)$$

Moreover, the scalar field satisfies the following equation of motion in the Friedmann universe:

$$\ddot{\phi} + 3H\dot{\phi} + V_\phi = 0, \qquad (8.8)$$

where $V_\phi = (dV/d\phi)$. From the above expressions for ρ_ϕ and p_ϕ, one finds that the condition for inflation, viz. $(\rho + 3p) < 0$, reduces to

$$\dot{\phi}^2 < V(\phi). \qquad (8.9)$$

In other words, inflation can be achieved if the potential energy of the scalar field dominates its kinetic energy.

Given a $V(\phi)$ that is motivated by a high energy model, the first of the Friedmann equations (8.5) and the equation (8.8) that governs the evolution of the scalar field have to be consistently solved for the scale factor and the scalar field, with suitable initial conditions. But, upon using the expressions (8.7) for the energy density and the pressure associated with the scalar field, the Friedmann equations (8.5) can be rewritten as

$$H^2 = \left(\frac{1}{3\,M_{\mathrm{P}}^2}\right)\left[\left(\frac{\dot{\phi}^2}{2}\right) + V(\phi)\right] \quad \text{and} \quad \dot{H} = -\left(\frac{1}{2\,M_{\mathrm{P}}^2}\right)\dot{\phi}^2, \quad (8.10)$$

where, for convenience, I have set $(8\,\pi\,G) = M_{\mathrm{P}}^{-2}$, as I had defined earlier. These two equations can then be combined to express the scalar field and the potential parametrically in terms of the cosmic time t as follows [1, 2]:

$$\phi(t) = \sqrt{2}\,M_{\mathrm{P}} \int dt \,\sqrt{\left(-\dot{H}\right)} \quad \text{and} \quad V(t) = M_{\mathrm{P}}^2 \left(3\,H^2 + \dot{H}\right). \quad (8.11)$$

If we know the scale factor $a(t)$, these two equations allow us to 'reconstruct' the potential from which such a scale factor can arise.

8.2.3   *Slow roll inflation*

The condition (8.9) that the potential energy of the inflaton dominates the kinetic energy is necessary for inflation to take place. However, inflation is *guaranteed*, if the field *rolls slowly* down the potential such that

$$\dot{\phi}^2 \ll V(\phi). \quad (8.12)$$

Moreover, it can be ensured that the field is slowly rolling for a sufficiently long time (to achieve the required 60 or so *e*-folds of inflation), provided

$$\ddot{\phi} \ll \left(3\,H\,\dot{\phi}\right). \quad (8.13)$$

These two conditions lead to the slow roll approximation [10], which, as we shall see, allows one to construct analytical solutions, both for the background and the perturbations. The approximation is usually described in terms of what are referred to as the slow roll parameters.

It can be readily shown that the two conditions (8.12) and (8.13) correspond to requiring the following two dimensionless parameters [10]:

$$\epsilon_{\mathrm{V}} = \left(\frac{M_{\mathrm{P}}^2}{2}\right)\left(\frac{V_\phi}{V}\right)^2 \quad \text{and} \quad \eta_{\mathrm{V}} = M_{\mathrm{P}}^2 \left(\frac{V_{\phi\phi}}{V}\right), \quad (8.14)$$

where $V_{\phi\phi} \equiv \left(\mathrm{d}^2 V/\mathrm{d}\phi^2\right)$, to be small when compared to unity. Given a potential $V(\phi)$, the two parameters ϵ_V and η_V—known as the Potential Slow Roll (PSR) parameters—immediately allow us to determine the domains and the parameters of the potential that can lead to inflation. However, it should be emphasized that the smallness of the PSR parameters alone cannot ensure slow roll inflation, since $\dot{\phi}$ can possibly be large. In addition to $(\epsilon_V, \eta_V) \ll 1$, the slow roll approximation actually requires that the scalar field is moving slowly along the attractor solution determined by the equation: $(3 H \dot{\phi}) \simeq -V_\phi$ [10]. A set of parameters that reflect the criteria (8.12) and (8.13) more accurately are the Hubble Slow Roll (HSR) parameters, which are defined by treating the Hubble parameter H as a function of the scalar field ϕ [10]. I shall work with these parameters in my discussion below.

When one considers the Hubble parameter H to be a function of the scalar field, we can write the second equation in Eq. (8.10) as

$$\dot{\phi} = -\left(2 M_P^2\right) H_\phi, \qquad (8.15)$$

where $H_\phi \equiv (\mathrm{d}H/\mathrm{d}\phi)$. This expression can then be used to rewrite the first Friedmann equation in Eq. (8.10) as follows:

$$H_\phi^2 - \left(\frac{3 H^2}{2 M_P^2}\right) = -\left(\frac{V}{2 M_P^4}\right), \qquad (8.16)$$

a relation that is referred to as the Hamilton-Jacobi formulation of inflation [10]. Taking $H(\phi)$ to be the primary quantity, the dimensionless HSR parameters ϵ_H and δ_H are defined as follows [10]:

$$\epsilon_H = \left(2 M_P^2\right) \left(\frac{H_\phi}{H}\right)^2 \quad \text{and} \quad \delta_H = \left(2 M_P^2\right) \left(\frac{H_{\phi\phi}}{H}\right), \qquad (8.17)$$

where $H_{\phi\phi} \equiv \left(\mathrm{d}^2 H/\mathrm{d}\phi^2\right)$. On using Eqs. (8.8), (8.15) and (8.16), these two parameters can be written as

$$\epsilon_H = \left(\frac{3 \dot{\phi}^2}{2 \rho_\phi}\right) = -\left(\frac{\dot{H}}{H^2}\right) \quad \text{and} \quad \delta_H = -\left(\frac{\ddot{\phi}}{H\dot{\phi}}\right) = \epsilon_H - \left(\frac{\dot{\epsilon}_H}{2 H \epsilon_H}\right), \qquad (8.18)$$

where ρ_ϕ is the energy density associated with the scalar field. It is clear from these expressions that, firstly, $\epsilon_H \ll 1$ is precisely the condition required for neglecting the kinetic energy term in the total energy of the scalar field. Secondly, the limit $\delta_H \ll 1$ corresponds to the situation wherein the acceleration term of the scalar field can be ignored in Eq. (8.8) when compared to the term involving the velocity. And, lastly, the inflationary condition $\ddot{a} > 0$ *exactly* corresponds to $\epsilon_H < 1$.

8.2.4   *Solutions in the slow roll approximation*

Note that the first Friedmann equation (8.10) and the equation of motion of the scalar field (8.8) can be written in terms of the two HSR parameters as

$$H^2 \left[1 - \left(\frac{\epsilon_{\text{H}}}{3}\right)\right] = \left(\frac{V}{3\,M_{\text{P}}^2}\right) \quad \text{and} \quad \left(3\,H\,\dot{\phi}\right) \left[1 - \left(\frac{\delta_{\text{H}}}{3}\right)\right] = -V_\phi.$$

$$(8.19)$$

The slow roll approximation corresponds to the situation wherein the HSR parameters ϵ_{H} and δ_{H} satisfy the following conditions:

$$\epsilon_{\text{H}} \ll 1, \quad \delta_{\text{H}} \ll 1, \quad \text{and} \quad \mathcal{O}\left[\epsilon_{\text{H}}^2, \delta_{\text{H}}^2, (\epsilon_{\text{H}} \delta_{\text{H}})\right] \ll \epsilon_{\text{H}}. \quad (8.20)$$

At the leading order in the slow roll approximation, the equations (8.19) above reduce to

$$H^2 \simeq \left(\frac{V}{3\,M_{\text{P}}^2}\right) \quad \text{and} \quad \left(3\,H\,\dot{\phi}\right) \simeq -V_\phi. \quad (8.21)$$

Being first order differential equations, given a potential, these equations can be readily integrated to obtain the solutions to the scale factor and the scalar field in the slow roll limit. As an illustration, let me now discuss the solutions in this limit to a certain class of potentials that are referred to as the large field models.

Consider potentials of the form [11]

$$V(\phi) = (V_0\,\phi^n), \quad (8.22)$$

where V_0 is a constant and $n > 0$. Let us restrict ourselves to the region $\phi > 0$ wherein $V(\phi)$ is positive for all n. Clearly, for such potentials, the parameters ϵ_{V} and η_{V} will be much less than unity provided $\phi \gg M_{\text{P}}$. Hence, inflation will occur in such potentials for large values of the field and, for this reason, these potentials are termed as 'large field' models [1, 2]. Moreover, inflation ends when, say, $\epsilon_{\text{V}} \simeq 1$, which corresponds to $\phi_{\text{end}} \simeq \left[(n/\sqrt{2})\,M_{\text{P}}\right]$ in these cases. For the above potential, when $n \neq 4$, upon integrating the first order equations (8.21), the solution to the scalar field in the slow roll limit is given by [1, 2]

$$\phi^{[(4-n)/2]}(t) \simeq \phi_{\text{i}}^{[(4-n)/2]} + \sqrt{\frac{V_0}{3}}\left[\frac{n\,(n-4)}{2}\right] M_{\text{P}}\,(t - t_{\text{i}}), \quad (8.23)$$

whereas, when $n = 4$, one finds that

$$\phi(t) \simeq \phi_{\text{i}}\,\exp-\left[\sqrt{(V_0/3)}\,(4\,M_{\text{P}})\,(t - t_{\text{i}})\right] \quad (8.24)$$

and, in both these solutions, ϕ_i is a constant that denotes the value of the scalar field at some initial time t_i. For all n, the scale factor can be expressed in terms of these solutions for the scalar field as follows:

$$a(t) \simeq a_i \exp - \left[\left(\frac{1}{2\,n\,M_P^2} \right) \left(\phi^2(t) - \phi_i^2 \right) \right] \qquad (8.25)$$

with a_i being the value of the scalar factor at t_i. It is also useful to note that, in the slow roll limit, the two equations (8.21) allow us to express the number of e-folds from t_i to t during inflation as

$$N = \ln \left(\frac{a}{a_i} \right) = \int_{t_i}^{t} \mathrm{d}t \; H \simeq - \left(\frac{1}{M_P^2} \right) \int_{\phi_i}^{\phi} \mathrm{d}\phi \left(\frac{V}{V_\phi} \right), \qquad (8.26)$$

where the upper limit ϕ is the value of the scalar field at the time t. In terms of e-folds, for the large field models, the scalar field and the Hubble parameter are given by

$$\phi^2(N) \simeq \left[\phi_i^2 - \left(2\,M_P^2\,n \right)\,N \right], \qquad (8.27a)$$

$$H^2(N) \simeq \left(\frac{V_0\,M_P^{(n-2)}}{3} \right) \left[\left(\frac{\phi_i}{M_P} \right)^2 - (2\,n\,N) \right]^{(n/2)}. \qquad (8.27b)$$

8.3   Gauge invariant, linear, perturbation theory

CMBR observations unambiguously point to the fact that the anisotropies at the epoch of decoupling are rather small (one part in 10^5, as I had mentioned). If so, the amplitude of the deviations from homogeneity will be even smaller at earlier epochs. This suggests that the generation and the evolution of the perturbations (until structures begin to form late in the matter dominated epoch) can be studied using linear perturbation theory.

In a Friedmann background, the metric perturbations can be decomposed according to their behavior under local rotation of the spatial coordinates on hypersurfaces of constant time. This property leads to the classification of the perturbations as scalars, vectors and tensors [12, 13]. Scalar perturbations remain invariant under rotations (and, hence, can be said to have zero spin), and they are the primary perturbations that are responsible for the inhomogeneities and the anisotropies in the universe. Vector and tensor perturbations—as their names indicate—transform as vectors and tensors do under rotations (and, as a result, have spins of one and two, respectively). The vector perturbations are generated by rotational velocity fields and, therefore, are also referred to as the vorticity modes. Finally, the tensor perturbations describe gravitational waves, and it is important to note that they can exist even in the absence of sources [14].

A simple counting immediately indicates that, when the freedom associated with the coordinate transformations (i.e. the gauge degrees of freedom) are eliminated, in $(3+1)$-dimensions, there exist two independent degrees of freedom each of the scalar, the vector and the tensor perturbations (for a detailed discussion on this point, see, for instance, the last but one reference in Refs. [2]). Moreover, it can be shown that, at the linear order, the scalar, vector and tensor perturbations evolve independently (i.e. they decouple), and it is therefore possible to analyze them separately [1, 2]. While considering the perturbations, one either works in a specific gauge to describe them, or works in a gauge invariant fashion, and I shall adopt the latter approach here. In what follows, I shall arrive at the equations of motion governing the gauge invariant variables that describe the scalar, the vector and the tensor degrees of freedom. In the case of the scalar perturbations, due to its importance in, say, understanding the effects on the CMBR, I shall also discuss the evolution of the scalar perturbations in the super-Hubble limit in some detail.

8.3.1 *Scalar perturbations*

If we now take into account the scalar perturbations to the background metric (8.1), then the Friedmann line-element, in general, can be written as [1, 2]

$$
\begin{aligned}
ds^2 = (1 + 2\,A)\ dt^2 &- 2\,a(t)\,(\partial_i B)\ dt\ dx^i \\
&- a^2(t)\ [(1 - 2\,\psi)\,\delta_{ij} + 2\,(\partial_i\,\partial_j E)]\ dx^i\,dx^j, \quad (8.28)
\end{aligned}
$$

where A, B, ψ and E are four scalar functions that describe the perturbations, which depend on time as well as space. However, recall that, I had mentioned above that there exist only two independent degrees of freedom describing the scalar perturbations. The two additional degrees of freedom arise due to the following scalar, gauge (i.e. infinitesimal coordinate) transformations that are possible [1, 2]

$$
t \to (t + \delta t) \quad \text{and} \quad x_i \to [x_i + \partial_i\,(\delta x)]\,. \quad (8.29)
$$

where δt and δx are scalar quantities that are functions of time and space. Clearly, the metric perturbations will not be invariant under such a change of coordinates and, it is easy to show that, under the gauge transformations (8.29), the functions A, B, ψ and E transform as follows:

$$A \to \left(A - \dot{\delta t}\right), \qquad B \to \left[B + (\delta t/a) - a\,\dot{\delta x}\right],$$

$$\psi \to (\psi + H\,\delta t), \qquad E \to (E - \delta x). \qquad (8.30)$$

At this stage, one could choose specific forms for the quantities δt and δx, thereby restricting oneself to a particular gauge. Two popular and convenient gauges that are often chosen to describe the scalar perturbations are the Newtonian gauge (wherein $B = E = 0$) and the uniform curvature (or the spatially flat) gauge (wherein $\psi = E = 0$). The other option would be to work in an explicitly gauge invariant fashion and, as I had mentioned earlier, it is this method that I shall adopt in our discussion below. The gauge invariant variables—known in literature as the Bardeen potentials [12]—that characterize the two degrees of freedom of the scalar perturbations are given by

$$\Phi = A + \left[a\left(B - a\,\dot{E}\right)\right]^{\cdot} \quad \text{and} \quad \Psi = \psi - \left[a\,H\left(B - a\,\dot{E}\right)\right]. \qquad (8.31)$$

On using the transformations (8.30) of the metric functions A, B, ψ and E, it is straightforward to check that these Bardeen potentials Φ and Ψ are indeed gauge invariant.

8.3.1.1 *Equations of motion governing the scalar perturbations*

At the linear order in the perturbations, the components of the perturbed Einstein tensor, viz. δG^{μ}_{ν}, corresponding to the line-element (8.28) are found to be

$$\delta G^0_0 = -6\,H\left(\dot{\psi} + H\,A\right) + (2/a^2)\,\nabla^2\left(\psi - \left[a\,H\left(B - a\,\dot{E}\right)\right]\right), \qquad (8.32a)$$

$$\delta G^0_i = 2\,\partial_i\left(\dot{\psi} + H\,A\right), \qquad (8.32b)$$

$$\delta G^i_j = -\left[2\,\ddot{\psi} + 2\,H\left(3\,\dot{\psi} + \dot{A}\right) + 2\left(2\,\dot{H} + 3\,H^2\right)A\right.$$

$$\left. + (1/a^2)\,\nabla^2\left((A - \psi) + (1/a)\left[a^2\left(B - a\,\dot{E}\right)\right]^{\cdot}\right)\right]\delta^i_j$$

$$+ (1/a^2)\,\partial^i\partial_j\left((A - \psi) + (1/a)\left[a^2\left(B - a\,\dot{E}\right)\right]^{\cdot}\right), \qquad (8.32c)$$

where $\nabla^2 = \left(\partial_i\,\partial^i\right)$. Under the gauge transformations (8.29), the components of the perturbed Einstein tensor δG^{μ}_{ν} transform as follows:

$$\delta G^0_0 \to \left(\delta G^0_0 - \dot{G}^0_0\,\delta t\right), \quad \delta G^0_i \to \delta G^0_i - \left[G^0_0 - (1/3)\,G^l_l\right](\partial_i\,\delta t), \qquad (8.33a)$$

$$\delta G^i_j \to \left(\delta G^i_j - \dot{G}^i_j\,\delta t\right), \qquad (8.33b)$$

where G^μ_ν denotes the Einstein tensor corresponding to the background metric, i.e. when the perturbations are absent. As in the case of the scalar functions describing the metric perturbations, it is straightforward to construct gauge invariant quantities corresponding to the perturbed Einstein tensor δG^μ_ν. The gauge invariant, perturbed Einstein tensor, which I shall denote as $\delta \mathcal{G}^\mu_\nu$, can be shown to be [1, 2]

$$\delta \mathcal{G}^0_0 = \delta G^0_0 + \dot{G}^0_0 \left[a \left(B - a \dot{E} \right) \right], \qquad (8.34\text{a})$$

$$\delta \mathcal{G}^0_i = \delta G^0_i + \left[G^0_0 - (1/3) \, G^l_l \right] \partial_i \left[a \left(B - a \dot{E} \right) \right], \qquad (8.34\text{b})$$

$$\delta \mathcal{G}^i_j = \delta G^i_j + \dot{G}^i_j \left[a \left(B - a \dot{E} \right) \right]. \qquad (8.34\text{c})$$

Let me now turn to the stress-energy tensor that describes the source of the perturbations. The only sources that I shall consider in this article will be scalar fields and vorticity free perfect fluids. As I shall illustrate in the next section, scalar fields do not possess any anisotropic stress at the linear order in the perturbations. I shall assume that the perfect fluids that I consider do not contain any anisotropic stresses either. Under these conditions, the perturbed stress-energy tensor associated with such sources can be expressed as follows:

$$\delta T^0_0 = \delta\rho, \quad \delta T^0_i = (\partial_i \, \delta\sigma) \quad \text{and} \quad \delta T^i_j = -\delta p \, \delta^i_j, \qquad (8.35)$$

where the quantities $\delta\rho$, $\delta\sigma$, and δp are the scalar quantities that denote the perturbations in the energy density, the energy flux, and the pressure, respectively. Evidently, the gauge invariant perturbed stress-energy tensor, which I shall denote as $\delta \mathcal{T}^\mu_\nu$, can be constructed in the same fashion as we had constructed the gauge invariant Einstein tensor $\delta \mathcal{G}^\mu_\nu$ above [cf. Eqs. (8.34)].

In the absence of anisotropic stresses, it can be readily shown that the non-diagonal, spatial components of the first order Einstein's equations, viz. $\delta \mathcal{G}^\mu_\nu = (8\pi G) \, \delta \mathcal{T}^\mu_\nu$, lead to the relation: $\Phi = \Psi$. The remaining first order Einstein's equations then reduce to [1, 2]

$$\left(\frac{1}{a^2} \right) \nabla^2 \Phi - 3 H \left(\dot{\Phi} + H \, \Phi \right) = (4\pi G) \left[\delta\rho + (\dot{\rho}\, a) \left(B - a \dot{E} \right) \right]$$

$$= (4\pi G) \, \delta\varrho, \qquad (8.36\text{a})$$

$$\partial_i \left(\dot{\Phi} + H \, \Phi \right) = (4\pi G) \, \partial_i \left[\delta\sigma + [(\rho + p)\, a] \left(B - a \dot{E} \right) \right]$$

$$= (4\pi G) \, (\partial_i \, \delta\varsigma), \qquad (8.36\text{b})$$

$$\ddot{\Phi} + 4 H \dot{\Phi} + \left(2\dot{H} + 3 H^2 \right) \Phi = (4\pi G) \left[\delta p + (\dot{p}\, a) \left(B - a \dot{E} \right) \right]$$

$$= (4\pi G) \, \delta\mathcal{P}, \qquad (8.36\text{c})$$

where $\delta\varrho$, $\delta\varsigma$ and $\delta\mathcal{P}$ represent the gauge invariant perturbations in the energy density, the energy flux, and the pressure of the matter field, respectively. The first and the third of the above first order Einstein's equations can be combined to lead to the following differential equation for the Bardeen potential Φ [1, 2]:

$$\Phi'' + 3\,\mathcal{H}\,\left(1 + c_{\text{A}}^2\right)\,\Phi' - c_{\text{A}}^2\,\nabla^2\Phi$$
$$+ \left[2\,\mathcal{H}' + \left(1 + 3\,c_{\text{A}}^2\right)\,\mathcal{H}^2\right]\,\Phi = \left(4\,\pi\,G\,a^2\right)\,\delta p^{\text{NA}}, \quad (8.37)$$

where $\mathcal{H} = (H\,a)$ is the conformal Hubble parameter. In arriving at this equation, I have changed over to the conformal time coordinate η, and have made use of the following standard definition of the non-adiabatic pressure perturbation δp^{NA} [15, 16]:

$$\delta p^{\text{NA}} = \left(\delta\mathcal{P} - c_{\text{A}}^2\,\delta\varrho\right) = \left(\delta p - c_{\text{A}}^2\,\delta\rho\right), \quad (8.38)$$

where the quantity $c_{\text{A}}^2 \equiv (\dot{p}/\dot{\rho})$ is often referred to as the adiabatic speed of the perturbations.

8.3.1.2 *A conserved quantity at super-Hubble scales*

Consider the following linear combination of the Bardeen potential Ψ and the gauge invariant energy flux $\delta\varsigma$ [17]:

$$\mathcal{R} = \Psi + \left(\frac{H}{\rho + p}\right)\,\delta\varsigma, \quad (8.39)$$

a quantity that is referred to as the curvature perturbation[6]. In the absence of anisotropic stress, upon using the second of the first order Einstein's equations (8.36), we can express the quantity $\mathcal{R}$ defined above as

$$\mathcal{R} = \Phi + \left(\frac{2\,\rho}{3\,\mathcal{H}}\right)\,\left(\frac{\Phi' + \mathcal{H}\,\Phi}{\rho + p}\right). \quad (8.40)$$

Upon substituting this expression in Eq. (8.37) that describes the evolution of the potential Φ, and making use of the background equations (8.5), one obtains that, in Fourier space,

$$\mathcal{R}_k' = \left(\frac{\mathcal{H}}{\mathcal{H}^2 - \mathcal{H}'}\right)\,\left[\left(4\,\pi\,G\,a^2\right)\,\delta p_k^{\text{NA}} - c_{\text{A}}^2\,k^2\,\Phi_k\right], \quad (8.41)$$

where, note that, the sub-scripts k refer to the wavenumber of the Fourier modes of the perturbations. Now, at super-Hubble scales, wherein the

---

[6]It is called so since it turns out to be proportional to the local three curvature on the spatial hypersurface [1, 2].

physical wavelengths of the perturbations are much larger than the Hubble radius [i.e. when $(k/a\,H) = (k/\mathcal{H}) \ll 1$], the term $\left(c_{\mathrm{A}}^2\,k^2\,\Phi_k\right)$ can be neglected. If one further assumes that no non-adiabatic pressure perturbations are present (i.e. $\delta p^{\mathrm{NA}} = 0$), say, as in the case of perfect, barotropic fluids [16], then the above equation implies that $\mathcal{R}'_k \simeq 0$ at super-Hubble scales. In other words, when the perturbations are adiabatic, the curvature perturbation $\mathcal{R}_k$ is conserved when the modes are well outside the Hubble radius.

8.3.1.3 *Evolution of the Bardeen potential at super-Hubble scales*

Let us now make use of the conservation of curvature perturbation to understand how the Bardeen potential evolves at super-Hubble scales during the radiation and matter dominated eras. If I now define [1, 2]

$$\Phi = \left(\mathcal{H}/a^2\,\theta\right)\mathcal{U}, \quad \text{where} \quad \theta = \left[\frac{\mathcal{H}^2}{(\mathcal{H}^2 - \mathcal{H}')\,a^2}\right]^{1/2}, \qquad (8.42)$$

then, I find that, in Fourier space, the equation (8.37) that governs the Bardeen potential reduces to

$$\mathcal{U}_k'' + \left[c_{\mathrm{A}}^2\,k^2 - \left(\frac{\theta''}{\theta}\right)\right]\mathcal{U}_k = \left(\frac{4\,\pi\,G\,a^4\,\theta}{\mathcal{H}}\right)\delta p_k^{\mathrm{NA}}. \qquad (8.43)$$

In the absence of non-adiabatic pressure perturbations (i.e. when $\delta p^{\mathrm{NA}} = 0$), the above differential equation for $\mathcal{U}_k$ simplifies to

$$\mathcal{U}_k'' + \left[c_{\mathrm{A}}^2\,k^2 - \left(\frac{\theta''}{\theta}\right)\right]\mathcal{U}_k = 0. \qquad (8.44)$$

And, in the super-Hubble limit, i.e. as $k \to 0$, the general solution to this differential equation can be written as [1, 2]

$$\mathcal{U}_k(\eta) \simeq C_{\mathrm{G}}(k)\,\theta(\eta)\int^\eta \frac{\mathrm{d}\tilde{\eta}}{\theta^2(\tilde{\eta})} + C_{\mathrm{D}}(k)\,\theta(\eta), \qquad (8.45)$$

where the coefficients C_{G} and C_{D} are k-dependent constants that are determined by the initial conditions imposed at early times. The corresponding Bardeen potential Φ_k is then given by

$$\Phi_k(\eta) \simeq C_{\mathrm{G}}(k)\left(\frac{\mathcal{H}}{a^2(\eta)}\right)\int^\eta \frac{\mathrm{d}\tilde{\eta}}{\theta^2(\tilde{\eta})} + C_{\mathrm{D}}(k)\left(\frac{\mathcal{H}}{a^2(\eta)}\right). \qquad (8.46)$$

Consider the following power law expansion:

$$a(t) = a_0\,t^q, \qquad (8.47)$$

which, for suitable values of the index q, can describe the radiation and matter dominated epochs as well as power law inflation. Such an expansion can be expressed in terms of the conformal time coordinate as follows:

$$a(\eta) = (-\bar{\mathcal{H}}\eta)^{(\gamma+1)}, \tag{8.48}$$

where γ and $\bar{\mathcal{H}}$ are constants given by

$$\gamma = -\left(\frac{2q-1}{q-1}\right) \quad \text{and} \quad \bar{\mathcal{H}} = \left[(q-1)a_0^{1/q}\right]. \tag{8.49}$$

During inflation (i.e. when $q > 1$), $\gamma \le -2$, with $\gamma = -2$ corresponding to exponential inflation (i.e. $q \to \infty$). While $\gamma = 0$ [corresponding to $q = (1/2)$] in the case of the radiation dominated epoch, $\gamma = 1$ [which corresponds to $q = (2/3)$] during matter domination. Moreover, note that, the quantity $\bar{\mathcal{H}}$ is positive and $-\infty < \eta < 0$ during inflation, whereas, during radiation and matter domination, $\bar{\mathcal{H}}$ is negative and $0 < \eta < \infty$. It is helpful to notice that, in all these instances, $\eta \to 0$ corresponds to the super-Hubble limit.

In the background (8.48), the quantity θ that is defined in Eq. (8.42) is found to be

$$\theta(\eta) = \left(\frac{\gamma+1}{\gamma+2}\right)^{1/2} \left(\frac{1}{a(\eta)}\right), \tag{8.50}$$

so that, on super-Hubble scales, one has [cf. Eq. (8.46)]

$$\Phi_k(\eta) \simeq C_{\mathrm{G}}(k) \left[\frac{3(w+1)}{3w+5}\right] + C_{\mathrm{D}}(k) \left[\frac{2/(3w+1)}{\bar{\mathcal{H}}^{[2(\gamma+1)]}\,\eta^{(2\gamma+3)}}\right], \tag{8.51}$$

where w is the following equation of state parameter:

$$w \equiv (p/\rho) = [(1-\gamma)/3(1+\gamma)] \tag{8.52}$$

that is a constant in power law expansion. The first term in the above expression for Φ_k denotes the growing mode (which is actually a constant), while the second term represents the decaying mode. (Hence, the choice of sub-scripts G and D to the coefficient C.) Demanding finiteness at very early times implies that the decaying mode has to be neglected, so that, at super-Hubble scales, I have [1, 2]

$$\Phi_k(\eta) \simeq C_{\mathrm{G}}(k) \left[\frac{3(w+1)}{3w+5}\right]. \tag{8.53}$$

This quantity vanishes when $w = -1$, which corresponds to exponential expansion driven by the cosmological constant. For this reason, it is often said that the cosmological constant does not induce any metric perturbations.

Since Φ_k is a constant at super-Hubble scales, in this limit, the curvature perturbation $\mathcal{R}_k$ [cf. Eq. (8.40)] is given by

$$\mathcal{R}_k \simeq \left[\frac{3\,w+5}{3\,(w+1)}\right]\Phi_k \simeq C_{\rm G}(k), \tag{8.54}$$

where I have made use of the expression (8.53) for Φ_k. As $\mathcal{R}_k$ is conserved and Φ_k is a constant at super-Hubble scales in power law expansion, when the modes enter the Hubble radius during the radiation or the matter dominated epochs, the Bardeen potentials at entry are given by [1, 2]

$$\Phi_k^{\rm R} \simeq \left[\frac{3\,(w_{\rm R}+1)}{3\,w_{\rm R}+5}\right]\mathcal{R}_k = \left(\frac{2}{3}\right)C_{\rm G}(k), \tag{8.55a}$$

$$\Phi_k^{\rm M} \simeq \left[\frac{3\,(w_{\rm M}+1)}{3\,w_{\rm M}+5}\right]\mathcal{R}_k = \left(\frac{3}{5}\right)C_{\rm G}(k), \tag{8.55b}$$

where I have made use of the fact that $w_{\rm R} = (1/3)$ and $w_{\rm M} = 0$. These expressions also imply that, at super-Hubble scales, Φ_k changes by a factor of $(9/10)$ during the transition from the radiation to the matter dominated epoch [1, 2].

It is essentially the spectrum of the Bardeen potential when the modes enter the Hubble radius during the radiation and the matter dominated epochs that determines the pattern of the anisotropies in the CMBR and the structure of the universe at the largest scales that we observe today [1]. The quantity $C_{\rm G}(k)$ that determines such a spectrum has to be arrived at by solving the equation (8.43) exactly with suitable initial conditions imposed in the early universe. In the inflationary scenario, physically well motivated, quantum, initial conditions are imposed on the modes when they are deep inside the Hubble radius. As I shall illustrate in the next section, under these conditions, it is the background dynamics during the inflationary regime that influences the functional form of $C_{\rm G}(k)$.

8.3.2   *Vector perturbations*

When the vector perturbations are included, the Friedmann metric is described by the line element [1]

$$ds^2 = dt^2 - 2\,a(t)\,S_i\,dt\,dx^i - a^2(t)\,[\delta_{ij} + (\partial_i\,F_j + \partial_j\,F_i)]\,dx^i\,dx^j, \tag{8.56}$$

and it has to be noted that the vectors S_i and F_i that represent the perturbations are divergence free. We had earlier pointed out that there exist two independent degrees of freedom associated with the vector perturbations in a Friedmann universe. But, the total number of degrees of freedom associated with the three dimensional, divergence free vectors S_i and F_i is actually four. The two extra degrees of freedom arise due to the possibility of vector gauge transformations of the following form:

$$t \to t \quad \text{and} \quad x_i \to (x_i + \delta x_i)\,, \tag{8.57}$$

where δx^i is a divergence free vector. In principle, the two degrees of freedom associated with δx^i can be utilized to restrict oneself to a specific gauge, say, wherein either S_i or F_i is zero. It is found that, under the gauge transformations (8.57), the quantities S_i or F_i transform as follows:

$$S_i \to \left(S_i - a\,\delta \dot{x}_i\right) \quad \text{and} \quad F_i \to (F_i - \delta x_i)\,. \tag{8.58}$$

A gauge invariant and divergence free quantity that describes the two degrees of freedom associated with the vector perturbations can be constructed to be

$$V_i = \left(S_i - a\,\dot{F}_i\right)\,. \tag{8.59}$$

Upon taking into account the divergence free conditions on S_i and F_i, the components of the perturbed Einstein tensor corresponding to the line-element (8.56) can be obtained to be

$$\delta G_0^0 = 0, \quad \delta G_i^0 = -\left(\frac{1}{2\,a}\right)\nabla^2 V_i, \tag{8.60a}$$

$$\delta G_j^i = \left(\frac{1}{2\,a}\right)\left[\partial^i\left(\dot{V}_j + 2\,H\,V_j\right) + \partial_j\left(\dot{V}^i + 2\,H\,V^i\right)\right]\,. \tag{8.60b}$$

And, since these components contain only V_i, clearly, they are already invariant under the gauge transformations (8.57). Therefore, in the case of vector perturbations, $\delta G_\nu^\mu = \delta \mathcal{G}_\nu^\mu$. I had said that we shall restrict our discussion to perturbations induced by scalar fields and vorticity free perfect fluids, both of which are pure scalar sources. In the absence of any vector sources, according to the first order Einstein's equations, the above non-zero components δG_i^0 and δG_j^i have to be equated to zero. These then immediately imply that the metric perturbation V_i vanishes identically. In other words, no vector perturbations are generated in the absence of sources with vorticity [1, 2].

8.3.3   *Tensor perturbations*

Let us now turn to the case of the tensor perturbations. Upon the inclusion of these perturbations, the Friedmann metric can be described by the line element [1, 2]

$$ds^2 = dt^2 - a^2(t)\,(\delta_{ij} + h_{ij})\,dx^i\,dx^j, \tag{8.61}$$

where h_{ij} is a symmetric, transverse and traceless quantity that characterizes the tensor perturbations. The transverse and traceless conditions (viz. $\partial_j h^{ij} = 0$ and $h_i^i = 0$) reduce the number of independent degrees of freedom of h_{ij} to two. Moreover, there are no remaining gauge degrees of freedom. The two independent degrees of freedom associated with h_{ij} correspond to the two types of polarization of the gravitational waves. It can be shown that, on imposing the transverse and the traceless conditions, the components of the perturbed Einstein tensor corresponding to the above line element simplify to [1, 2]

$$\delta G_0^0 = \delta G_i^0 = 0, \quad \delta G_j^i = -\left(\frac{1}{2}\right)\left[\ddot{h}_j^i + 3\,H\,\dot{h}_j^i - \left(\frac{1}{a^2}\right)\nabla^2 h_j^i\right]. \qquad (8.62)$$

In the absence of anisotropic stresses, one then arrives at the following differential equation describing the gravitational waves [1, 2]:

$$h_{ij}'' + 2\,\mathcal{H}\,h_{ij}' - \nabla^2 h_{ij} = 0, \qquad (8.63)$$

where, for later use, I have expressed the equation in terms of the conformal time coordinate.

8.4 Generation of perturbations during inflation

As I have pointed out repeatedly, the most attractive feature of inflation is the fact that it provides a natural mechanism to generate the perturbations [4]. It is the quantum fluctuations associated with the inflaton that act as the primordial seeds for the inhomogeneities. I had earlier alluded to the fact that it is the spectrum of the Bardeen potential that determines the pattern of the anisotropies in the CMBR and the formation of structures. Since the curvature perturbation is proportional to the Bardeen potential at super-Hubble scales [cf. Eq. (8.54)], the primary quantity of interest is the spectrum of curvature perturbations generated during inflation. The inflaton being a scalar source, it does not degenerate any vector perturbations [1]. However, as I have discussed, gravitational waves are generated even in the absence of sources [14]. The primordial gravitational waves are also important because they too leave their own distinct imprints on the CMBR [1]. In this section, I shall first obtain the equation of motion governing the curvature perturbation when the universe is dominated by the inflaton. I shall then quantize the curvature and the tensor perturbations, impose vacuum initial conditions on the Fourier modes when they are well inside the Hubble radius, and evaluate the scalar and the tensor spectra in the super-Hubble limit for the case of slow roll inflation.

8.4.1 *Equation of motion for the curvature perturbation*

As before, let ϕ denote the homogeneous scalar field. Also, let $\delta\phi$ represent the perturbation in the field. It is then straightforward to show that, in the metric (8.28), the components of the perturbed stress-energy tensor [cf. Eqs. (8.6) and (8.35)] associated with the scalar field can be expressed as

$$\delta T^0_0 = \left(\dot{\phi}\,\dot{\delta\phi} - \dot{\phi}^2\,A + V_\phi\,\delta\phi \right) = \delta\rho_\phi, \tag{8.64a}$$

$$\delta T^0_i = \partial_i \left(\dot{\phi}\,\delta\phi \right) = \partial_i \left(\delta\sigma_\phi \right), \tag{8.64b}$$

$$\delta T^i_j = - \left(\dot{\phi}\,\dot{\delta\phi} - \dot{\phi}^2\,A - V_\phi\,\delta\phi \right) \delta^i_j = -\delta p_\phi\,\delta^i_j. \tag{8.64c}$$

The gauge invariant version of this stress-energy tensor can be obtained to be

$$\delta\mathcal{T}^0_0 = \left(\dot{\phi}\,\dot{\delta\varphi} - \dot{\phi}^2\,\Phi + V_\phi\,\delta\varphi \right) = \delta\varrho_\phi, \tag{8.65a}$$

$$\delta\mathcal{T}^0_i = \partial_i \left(\dot{\phi}\,\delta\varphi \right) = \partial_i \left(\delta\varsigma_\phi \right), \tag{8.65b}$$

$$\delta\mathcal{T}^i_j = - \left(\dot{\phi}\,\dot{\delta\varphi} - \dot{\phi}^2\,\Phi - V_\phi\,\delta\varphi \right) \delta^i_j = -\delta\mathcal{P}_\phi\,\delta^i_j. \tag{8.65c}$$

where the quantity $\delta\varphi$ denotes the gauge invariant perturbation in the scalar field, and is given by [1, 2]

$$\delta\varphi = \left[\delta\phi + (\dot{\phi}\,a)\,(B - a\,\dot{E}) \right]. \tag{8.66}$$

Evidently, the scalar field does not possess any anisotropic stress. As a result, $\Phi = \Psi$ during inflation. On substituting the above expressions for $\delta\varrho$, $\delta\varsigma$, and $\delta\mathcal{P}$ in the first order Einstein's equations (8.36) governing the scalar perturbations, one can arrive at the following equation for the Bardeen potential:

$$\Phi'' + 3\,\mathcal{H}\left(1 + c^2_{\rm A} \right) \Phi' - c^2_{\rm A}\,\nabla^2\Phi + \left[2\,\mathcal{H}' + \left(1 + 3\,c^2_{\rm A} \right) \mathcal{H}^2 \right] \Phi = \left(1 - c^2_{\rm A} \right) \nabla^2\Phi. \tag{8.67}$$

Upon comparing this equation for Φ with the general equation (8.37), it is evident that the non-adiabatic pressure perturbation associated with the inflaton is given by [16]

$$\delta p^{\rm NA}_\phi = \left(\frac{1 - c^2_{\rm A}}{4\,\pi\,G\,a^2} \right) \nabla^2\Phi. \tag{8.68}$$

In such a case, the equation (8.41) that describes the evolution of the curvature perturbation simplifies to

$$\mathcal{R}'_k = - \left(\frac{\mathcal{H}}{\mathcal{H}^2 - \mathcal{H}'} \right) \left(k^2\,\Phi_k \right). \tag{8.69}$$

On differentiating this equation again with respect to time and, on using the background equations (8.10), the definition (8.40) and the Bardeen equation (8.67), one obtains the following equation of motion governing the Fourier modes of the curvature perturbation induced by the scalar field [1, 2]:

$$\mathcal{R}_k'' + 2\left(\frac{z'}{z}\right)\mathcal{R}_k' + k^2\,\mathcal{R}_k = 0, \tag{8.70}$$

where the quantity z is given by

$$z = \left(a\,\dot{\phi}/H\right) = (a\,\phi'/\mathcal{H})\,. \tag{8.71}$$

It is useful to introduce the Mukhanov-Sasaki variable v that is defined as [18]

$$v = (\mathcal{R}\,z)\,. \tag{8.72}$$

The Fourier modes of this variable, say, v_k, satisfy the differential equation

$$v_k'' + \left[k^2 - \left(\frac{z''}{z}\right)\right]v_k = 0\,. \tag{8.73}$$

8.4.2   *Quantization of the perturbations and the definition of the power spectra*

On quantization, the homogeneity of the Friedmann background allows us to express the curvature perturbation $\mathcal{R}$ in terms of the Fourier modes $\mathcal{R}_k$ [satisfying Eq. (8.70)] as follows [2]:

$$\hat{\mathcal{R}}\,(\eta,\mathbf{x}) = \int \frac{\mathrm{d}^3\mathbf{k}}{(2\,\pi)^{3/2}}\left[\hat{a}_{\mathbf{k}}\,\mathcal{R}_k(\eta)\,e^{i\,\mathbf{k}\cdot\mathbf{x}} + \hat{a}_{\mathbf{k}}^{\dagger}\,\mathcal{R}_k^{*}(\eta)\,e^{-i\,\mathbf{k}\cdot\mathbf{x}}\right], \tag{8.74}$$

where the annihilation and the creation operators $\hat{a}_{\mathbf{k}}$ and $\hat{a}_{\mathbf{k}}^{\dagger}$ obey the standard commutation relations. At the linear order in the perturbation theory that I am working in, the power spectrum as well as the statistical properties of the scalar perturbations are entirely characterized by the two point function of the quantum field $\hat{\mathcal{R}}$[7]. The power spectrum of the scalar perturbations, say, $\mathcal{P}_\mathrm{s}\,(k)$, is given by the relation

$$\int_{0}^{\infty}\frac{dk}{k}\,\mathcal{P}_\mathrm{s}(k) \equiv \int \mathrm{d}^3\mathbf{k} \int \frac{\mathrm{d}^3\,(\mathbf{x}-\mathbf{x}')}{(2\,\pi)^3}\,\langle 0|\hat{\mathcal{R}}(\eta,\mathbf{x})\,\hat{\mathcal{R}}(\eta,\mathbf{x}')|0\rangle$$

$$\times \exp -i\,[\mathbf{k}\cdot(\mathbf{x}-\mathbf{x}')]\,, \tag{8.75}$$

---

[7]This is why the perturbations generated during inflation are usually referred to as Gaussian. However, this is essentially due to the fact that we are restricting ourselves to the linear order in perturbation theory. Deviations from Gaussianity can arise when one takes into account perturbations at the higher orders (see, for instance, Refs. [19], and references therein). As I had mentioned, I shall very briefly touch upon this point in the concluding section.

where $|0\rangle$ is the vacuum state defined as $\hat{a}_{\mathbf{k}}|0\rangle = 0 \ \forall \ \mathbf{k}$. Using the decomposition (8.74), the scalar perturbation spectrum can then be obtained to be [1, 2]

$$\mathcal{P}_{\mathrm{s}}(k) = \left(\frac{k^3}{2\pi^2}\right) |\mathcal{R}_k|^2 = \left(\frac{k^3}{2\pi^2}\right) \left(\frac{|v_k|}{z}\right)^2. \qquad (8.76)$$

The expression on the right hand side is to be evaluated at super-Hubble scales [i.e. when $(k/aH) = (k/\mathcal{H}) \ll 1$] when the curvature perturbation approaches a constant value[8].

The tensor perturbations can be quantized in a similar fashion as the curvature perturbation. Let us denote the amplitude of the tensor perturbations h_{ij} as h. Upon quantization, the amplitude h can be expressed in terms of the corresponding Fourier modes h_k as follows:

$$\hat{h}(\eta, \mathbf{x}) = \left(\frac{\sqrt{2}}{M_{\mathrm{P}}}\right) \int \frac{\mathrm{d}^3\mathbf{k}}{(2\pi)^{3/2}} \left[\hat{a}_{\mathbf{k}} \, h_k(\eta) \, e^{i\mathbf{k}\cdot\mathbf{x}} + \hat{a}_{\mathbf{k}}^\dagger \, h_k^*(\eta) \, e^{-i\mathbf{k}\cdot\mathbf{x}}\right], \quad (8.77)$$

where $\hat{a}_{\mathbf{k}}$ and $\hat{a}_{\mathbf{k}}^\dagger$ are the standard annihilation and creation operators as in the scalar case, while the additional factor of $(\sqrt{2}/M_{\mathrm{P}})$ arises because of the $(16\pi G)$ term in the Einstein-Hilbert action [1, 2]. The tensor power spectrum $\mathcal{P}_{\mathrm{T}}(k)$ is defined as

$$\int_0^\infty \frac{dk}{k} \, \mathcal{P}_{\mathrm{T}}(k) \equiv \int \mathrm{d}^3\mathbf{k} \int \frac{\mathrm{d}^3(\mathbf{x}-\mathbf{x}')}{(2\pi)^3} \sum_{i,j} \langle 0|\hat{h}_{ij}(\eta,\mathbf{x}) \, \hat{h}^{ij}(\eta,\mathbf{x}')|0\rangle$$
$$\times \exp{-i\left[\mathbf{k}\cdot(\mathbf{x}-\mathbf{x}')\right]},$$
$$= \int \mathrm{d}^3\mathbf{k} \int \frac{\mathrm{d}^3(\mathbf{x}-\mathbf{x}')}{(2\pi)^3} \, 2 \sum_{\lambda=(+,\times)} \langle 0|\hat{h}_\lambda(\eta,\mathbf{x}) \, \hat{h}_\lambda(\eta,\mathbf{x}')|0\rangle$$
$$\times \exp{-i\left[\mathbf{k}\cdot(\mathbf{x}-\mathbf{x}')\right]}, \quad (8.78)$$

where $|0\rangle$ again denotes the vacuum state as in the scalar case, and the sum in the final expression is over the two polarization states of the gravitational waves. If we write $h = (u/a)$, then, in Fourier space, the equation (8.63) describing the tensor perturbations reduces to

$$u_k'' + \left[k^2 - \left(\frac{a''}{a}\right)\right] u_k = 0. \qquad (8.79)$$

---

[8]Earlier, in Subsec. 8.3.1.2, I had illustrated that, if the non-adiabatic pressure perturbation δp^{NA} can be neglected, the curvature perturbation is conserved at super-Hubble scales. It can be shown that the non-adiabatic pressure perturbation associated with the inflaton [cf. Eq. (8.68)] decays exponentially at super-Hubble scales and, hence, can be ignored [20]. As a matter of fact, it is for this reason that the scalar perturbations produced by single scalar fields are usually referred to as adiabatic.

And, it is helpful to note that this equation is essentially the same as the Mukhanov-Sasaki equation (8.73) with the quantity z replaced by the scale factor a. The tensor perturbation spectrum can then be expressed in terms of the modes h_k and u_k as follows:

$$\mathcal{P}_{\mathrm{T}}(k) = \left(\frac{8}{M_{\mathrm{P}}^2}\right)\left(\frac{k^3}{2\pi^2}\right)|h_k|^2 = \left(\frac{8}{M_{\mathrm{P}}^2}\right)\left(\frac{k^3}{2\pi^2}\right)\left(\frac{|u_k|}{a}\right)^2 \qquad (8.80)$$

with the expressions on the right hand sides to be evaluated in the super-Hubble limit, as in case of the scalar perturbation spectrum.

The scalar and the tensor spectral indices are defined as [1]

$$n_{\mathrm{S}} = 1 + \left(\frac{d\ln\mathcal{P}_{\mathrm{S}}}{d\ln k}\right) \quad \text{and} \quad n_{\mathrm{T}} = \left(\frac{d\ln\mathcal{P}_{\mathrm{T}}}{d\ln k}\right). \qquad (8.81)$$

Notice the difference in the definition of these two quantities. Conventionally, a scale invariant scalar spectrum corresponds to $n_{\mathrm{S}} = 1$, while such a tensor spectrum is described by $n_{\mathrm{T}} = 0$. Finally, the tensor-to-scalar ratio r is defined as follows [1, 2]:

$$r(k) \equiv \left(\frac{\mathcal{P}_{\mathrm{T}}(k)}{\mathcal{P}_{\mathrm{S}}(k)}\right). \qquad (8.82)$$

As I shall briefly point out below, the scalar spectral index n_{S} and the tensor-to-scalar ratio r happen to be important inflationary parameters that can be constrained by the observations.

8.4.3 *The scalar and tensor spectra in slow roll inflation*

A large class of inflaton potentials admit a suitably long epoch of slow roll. So, let us now evaluate the scalar and tensor spectra in slow roll inflation. Using equations (8.10) and the expression for the first Hubble slow roll parameter ϵ_{H} in Eq. (8.18), the quantity z defined in equation (8.71) can be written as

$$z = \sqrt{2}\, M_{\mathrm{P}} \left(a\,\sqrt{\epsilon_{\mathrm{H}}}\right). \qquad (8.83)$$

Also, note that the equations (8.18) defining the two Hubble slow roll parameters ϵ_{H} and δ_{H} can be expressed in terms of the conformal time coordinate as follows:

$$\epsilon_{\mathrm{H}} = 1 - \left(\frac{\mathcal{H}'}{\mathcal{H}^2}\right) \quad \text{and} \quad \delta_{\mathrm{H}} = \epsilon_{\mathrm{H}} - \left(\frac{\epsilon_{\mathrm{H}}'}{2\,\mathcal{H}\,\epsilon_{\mathrm{H}}}\right). \qquad (8.84)$$

Using these expressions for z and the HSR parameters, the term (z''/z) that appears in the Mukhanov-Sasaki equation (8.73) can be written as [21]

$$\left(\frac{z''}{z}\right) = \mathcal{H}^2 \left[2 - \epsilon_{\mathrm{H}} + (\epsilon_{\mathrm{H}} - \delta_{\mathrm{H}})(3 - \delta_{\mathrm{H}}) + \left(\frac{\epsilon_{\mathrm{H}}' - \delta_{\mathrm{H}}'}{\mathcal{H}}\right)\right]. \qquad (8.85)$$

From the definition of ϵ_{H} above, it can also be established that

$$\left(\frac{a''}{a}\right) = \mathcal{H}^2 \left(2 - \epsilon_{\mathrm{H}}\right). \tag{8.86}$$

Let us now rewrite the expression (8.84) above for ϵ_{H} as follows:

$$\eta = -\int \left(\frac{1}{1 - \epsilon_{\mathrm{H}}}\right) \mathrm{d}\left(\frac{1}{\mathcal{H}}\right). \tag{8.87}$$

On integrating this expression by parts, and using the above definition of δ_{H}, one obtains that

$$\eta = -\left[\frac{1}{(1 - \epsilon_{\mathrm{H}})\,\mathcal{H}}\right] - \int \left[\frac{2\,\epsilon_{\mathrm{H}}\,(\epsilon_{\mathrm{H}} - \delta_{\mathrm{H}})}{(1 - \epsilon_{\mathrm{H}})^3}\right] \mathrm{d}\left(\frac{1}{\mathcal{H}}\right). \tag{8.88}$$

At the leading order in the slow roll approximation [cf. Eq. (8.20)], the second term can be ignored and, at the same order, one can assume ϵ_{H} to be a constant. Therefore, we have

$$\mathcal{H} \simeq -\left[\frac{1}{(1 - \epsilon_{\mathrm{H}})\,\eta}\right]. \tag{8.89}$$

If we now use this expression for $\mathcal{H}$ in the expressions (8.85) and (8.86), then, at the leading order in the slow roll approximation, one gets that

$$\left(\frac{z''}{z}\right) \simeq \left(\frac{2 + 6\,\epsilon_{\mathrm{H}} - 3\,\delta_{\mathrm{H}}}{\eta^2}\right) \quad \text{and} \quad \left(\frac{a''}{a}\right) \simeq \left(\frac{2 + 3\,\epsilon_{\mathrm{H}}}{\eta^2}\right), \tag{8.90}$$

with the slow roll parameters treated as constants.

As I have mentioned, during inflation, the initial conditions on the perturbations are imposed when the modes are well inside the Hubble radius [i.e. when $(k/a\,H) = (k/\mathcal{H}) \gg 1$]. It is clear from Eqs. (8.73) and (8.79) that, in such a sub-Hubble limit, the scalar and tensor modes v_k and u_k do not feel the curvature of the spacetime and, hence, the solutions to these modes behave in the following Minkowskian form: $e^{\pm(i\,k\,\eta)}$. The assumption that the scalar and tensor perturbations are in the vacuum state then requires that v_k and u_k are positive frequency modes at sub-Hubble scales, i.e. they have the asymptotic form [1, 2]

$$\lim_{(k/\mathcal{H}) \to \infty} (v_k(\eta), u_k(\eta)) \to \left(\frac{1}{\sqrt{2\,k}}\right) e^{-ik\eta}. \tag{8.91}$$

It should be pointed out that the vacuum state associated with the modes that exhibit such a behavior is often referred to in the literature as the Bunch-Davies vacuum [22].

For the quantity (z''/z) in Eq. (8.90), when the slow roll parameters are assumed to be constant, the solution to the Mukhanov-Sasaki equation (8.73) that satisfies the initial condition (8.91) can be obtained to be [23]

$$v_k(\eta) = \left(\frac{-\pi \eta}{4} \right)^{1/2} e^{i \, [\nu+(1/2)] \, (\pi/2)} \, H_\nu^{(1)} \, (-k\eta),$$ (8.92)

where $H_\nu^{(1)}$ is the Hankel function of the first kind and of order ν. Since, in the slow roll limit, the quantity (a''/a) depends on η in the same way as the corresponding (z''/z) does, it is evident that the tensor mode u_k [satisfying Eq. (8.79), and the initial conditions (8.91)] will be described by the same solution as the one for v_k above. The quantity ν is found to be

$$\nu_S \simeq \left[\left(\frac{3}{2} \right) + 2\,\epsilon_H - \delta_H \right] \quad \text{and} \quad \nu_T \simeq \left[\left(\frac{3}{2} \right) + \epsilon_H \right],$$ (8.93)

with the subscripts S and T referring to the scalar and tensor cases, respectively.

It now remains to evaluate the two spectra in the super-Hubble limit. In this limit [i.e. as $(-k\,\eta) \to 0$], upon expanding the Hankel function as a series about the origin, the scalar and the tensor spectra can be expressed as [2, 21]

$$\mathcal{P}_S(k) = \left(\frac{1}{32 \, \pi^2 \, M_P^2 \, \epsilon_H} \right) \left[\frac{|\Gamma(\nu_S)|}{\Gamma(3/2)} \right]^2 \left(\frac{k}{a} \right)^2 \left(\frac{-k\eta}{2} \right)^{(1-2\nu_S)}$$

$$= \left(\frac{H^2}{2\,\pi\,\dot\phi} \right)^2 \left[\frac{|\Gamma(\nu_S)|}{\Gamma(3/2)} \right]^2 2^{(2\nu_S-3)} \, (1 - \epsilon_H)^{(2\nu_S-1)},$$ (8.94a)

$$\mathcal{P}_T(k) = \left(\frac{1}{2 \, \pi^2 \, M_P^2} \right) \left[\frac{|\Gamma(\nu_T)|}{\Gamma(3/2)} \right]^2 \left(\frac{k}{a} \right)^2 \left(\frac{-k\eta}{2} \right)^{(1-2\nu_T)}$$

$$= \left(\frac{2\,H^2}{\pi^2 \, M_P^2} \right) \left[\frac{|\Gamma(\nu_T)|}{\Gamma(3/2)} \right]^2 2^{(2\nu_T-3)} \, (1 - \epsilon_H)^{(2\nu_T-1)},$$ (8.94b)

where H is the Hubble parameter, and the second equalities express the asymptotic values in terms of the values of the quantities at Hubble exit [i.e. at $(-k\,\eta) = (1-\epsilon_H)^{-1}$]. At the leading order in the slow roll approximation, the amplitudes of the scalar and the tensor spectra can easily be read off from the above expressions. They are given by [1, 2]

$$\mathcal{P}_S(k) \simeq \left(\frac{H^2}{2\,\pi\,\dot\phi} \right)^2_{k=(a\,H)} \quad \text{and} \quad \mathcal{P}_T(k) \simeq \left(\frac{8}{M_P^2} \right) \left(\frac{H}{2\,\pi} \right)^2_{k=(a\,H)},$$ (8.95)

with the sub-scripts on the right hand side indicating that the quantities have to be evaluated when the modes cross the Hubble radius. Given a quantity, say, y, we can write [2]

$$\left(\frac{dy}{d\ln k}\right)_{k=(aH)} = \left(\frac{dy}{dt}\right)\left(\frac{dt}{d\ln a}\right)\left(\frac{d\ln a}{d\ln k}\right)_{k=(aH)} = \left(\frac{\dot{y}}{H}\right)_{k=(aH)},$$

(8.96)

where, in arriving at the final expression, the following condition has been used:

$$\left(\frac{d\ln a}{d\ln k}\right)_{k=(aH)} \simeq 1,$$

(8.97)

as H does not vary much during slow roll inflation. Using the expressions (8.95) for the power spectra, the definitions (8.81) of the spectral indices and Eq. (8.96), one can easily show that

$$n_{\mathrm{S}} \simeq (1 - 4\,\epsilon_{\mathrm{H}} + 2\,\delta_{\mathrm{H}}) \quad \text{and} \quad n_{\mathrm{T}} \simeq -(2\,\epsilon_{\mathrm{H}}).$$

(8.98)

These expressions unambiguously point to the fact the scalar and the tensor spectra that arise in slow roll inflation will be nearly scale invariant. The tensor-to-scalar ratio in the slow roll limit is found to be

$$r \simeq (16\,\epsilon_{\mathrm{H}}) = -(8\,n_{\mathrm{T}})$$

(8.99)

with the last equality often referred to as the consistency relation [2].

Recent observations of the anisotropies in the CMBR seem to strongly favor a nearly scale invariant, scalar perturbation spectrum, as is predicted by slow roll inflation (see, for instance, Refs. [5]). The observations suggest a scalar spectral amplitude of about 2.137×10^{-9} at a pivot scale of about 0.05 Mpc^{-1} (a point that is usually referred to as COBE normalization; in this context, see Ref. [24]), and a scalar spectral index n_{S} of about 0.96 [5]. While tensors remain undetected, current data constrain the tensor-to-scalar ratio to be $r \lesssim 0.4$ [5]. It should be mentioned that, despite the extraordinary precision of the CMBR observations, at this stage, a plethora of inflationary models still remain consistent with the data.

8.5   Reheating the universe

As I had discussed in the introductory section, during inflation, it is assumed that radiation does not interact with the inflaton. The radiation, being free, its temperature behaves as inversely proportional to the scale factor [1]. Hence, the universe cools down by the same large factor, viz.

$A \simeq 10^{26}$, that it expands by during the inflationary epoch. Therefore, at the end of inflation, the universe needs to be quickly heated back to the same temperature as it was at the beginning of the epoch, if we are to return to the standard, early, radiation dominated phase of the hot big bang model. This is achieved through coupling, by hand, the inflaton to radiation at the end of inflation [7]. Such a coupling is expected to effectively capture, in a simple and phenomenological way, the decay of the inflaton into the various particles that constitute, say, the standard model [8]. In this section, we shall first consider the behavior of the scalar field at the end of inflation, in the absence of any coupling to radiation, and then discuss the transfer of the energy from the inflaton to radiation, achieved by coupling them through a coarse grained decay rate.

8.5.1 *Behavior of the scalar field at the end of inflation*

In virtually all the conventional models of inflation, the scalar field rolls towards a minima. Typically, inflation is terminated as the field approaches the minima and, thereafter, the field begins to oscillate about the minimum of the potential. Consider, for instance, the large field models (8.22) that we had discussed earlier. In such models, the scalar field rolls towards the minimum of the potential from large values, and inflation ends as the first slow roll parameter, say, ϵ_{v}, approaches unity around $\phi_{\mathrm{end}} = [(n/\sqrt{2}) \, M_{\mathrm{P}}]$. For even n, the field then oscillates at the bottom of the potential. During this regime, the system is governed by two time scales, viz. the Hubble time H^{-1} and the period of the oscillations around the minimum which is determined by the quantity $V_{\phi\phi}$. Importantly, these two scales prove to be rather different, with the frequency of the oscillations being much larger than the Hubble scale [2]. To understand the effects of the oscillations of the scalar field at the bottom of the potential on the background, let us now study the evolution of the time averaged energy of the scalar field.

Note that, the equation of motion (8.8) of the inflaton can be written as

$$\dot{\rho}_\phi = -\left(3 \, H \, \dot{\phi}^2\right) = -6 \, H \left[\rho_\phi - V(\phi)\right], \tag{8.100}$$

where ρ_ϕ is the energy density of the scalar field given by Eq. (8.7). Upon averaging this equation over a time period of oscillation of the inflaton at the bottom of the potential, we obtain that

$$\langle \dot{\rho}_\phi \rangle = -\langle 6 \, H \left[\rho_\phi - V(\phi)\right] \rangle \simeq -6 \, H \, \langle \rho_\phi - V(\phi) \rangle, \tag{8.101}$$

where the angular brackets denote the time averages and, in arriving at the final expression, we have made use of the fact that the Hubble rate does not change much over a period of the oscillation. We can now write

$$\langle \rho_\phi - V(\phi) \rangle \equiv \left(\frac{1}{\mathcal{T}} \right) \int_0^{\mathcal{T}} dt \, [\rho_\phi - V(\phi)]$$

$$= \left[\int_{-\phi_m}^{\phi_m} d\phi \, \sqrt{\rho_\phi - V(\phi)} \right] \left[\int_{-\phi_m}^{\phi_m} \frac{d\phi}{\sqrt{\rho_\phi - V(\phi)}} \right]^{-1} , \quad (8.102)$$

where $\mathcal{T}$ and ϕ_m are the time period and the value of the scalar field when it has reached the maximum amplitude during the oscillations. Also, we have converted the integral over time to an integral over the field using the relation: $dt = \left(d\phi / \sqrt{2 \, [\rho_\phi - V(\phi)]} \right)$. If we now assume that, over one period, $\rho_\phi \simeq V(\phi_m) = V_m$ is a constant, then we can easily carry out the integrals in the above expression for the case of the large field models (8.22) with even n, to obtain that

$$\langle \rho_\phi - V(\phi) \rangle \simeq (\alpha \, \rho_\phi), \quad (8.103)$$

where α is a number that can shown to be [7]

$$\alpha \equiv \left[\int_{-\phi_m}^{\phi_m} d\phi \, \sqrt{1 - [V(\phi)/V_m]} \right] \left[\int_{-\phi_m}^{\phi_m} \frac{d\phi}{\sqrt{1 - [V(\phi)/V_m]}} \right]^{-1} = \left(\frac{n}{n+2} \right).$$

$$(8.104)$$

In such a case, the equation (8.101) that governs the average value of the energy density of the scalar field reduces to

$$\dot{\rho}_\phi \simeq - (3 \, H \, \bar{\alpha}) \, \rho_\phi, \quad (8.105)$$

where, for convenience, we have dropped the angular brackets and we have set $\bar{\alpha} = (2 \, \alpha)$.

The above equation for ρ_ϕ can be easily integrated to arrive at the result [7]

$$\rho_\phi = \rho_o \, (a/a_o)^{-(3 \, \bar{\alpha})} = \rho_o \, (a/a_o)^{-[6 \, n/(n+2)]} , \quad (8.106)$$

where ρ_o denotes the energy density of the scalar field when the scale factor was a_o, say, at the time t_o when the scalar field begins to oscillate at the bottom of the inflaton potential. This implies that, when $n = 2$ (i.e. in a quadratic potential), the average energy density ρ_ϕ of a scalar field that is oscillating at the bottom of the potential behaves as pressureless, non-relativistic matter, whereas, when $n = 4$ (viz. in a quartic potential), it behaves as radiation.

8.5.2   *Transferring the energy from the inflation to radiation*

Until now, we have confined our discussion to situations wherein there was only one source present, i.e. either a perfect fluid or a scalar field. We now need to consider scenarios wherein both a scalar field and radiation are present. Moreover, they need to interact as well, if we are to achieve the transfer of energy from the scalar field to radiation. We have denoted the stress-energy tensor associated with the scalar field as $T^{\mu\nu}_{(\phi)}$ [cf. Eq. (8.6)]. Let $T^{\mu\nu}_{(\gamma)}$ denote the stress-energy tensor corresponding to radiation. If we now allow for a transfer of energy between the components, then the covariant derivatives of the stress-energy tensors of the two components will be given by

$$\nabla_\mu T^{\mu\nu}_{(\phi)} = Q^\nu_{(\phi)} \quad \text{and} \quad \nabla_\mu T^{\mu\nu}_{(\gamma)} = Q^\nu_{(\gamma)}, \tag{8.107}$$

where $Q^\nu_{(\phi)}$ and $Q^\nu_{(\gamma)}$ are four vectors that represent the transfer of energy-momentum from radiation to the inflaton and vice-versa. Since the total stress-energy tensor, viz. $T^{\mu\nu} = [T^{\mu\nu}_{(\phi)} + T^{\mu\nu}_{(\gamma)}]$, has to be conserved, we also require that

$$\nabla_\mu \left[T^{\mu\nu}_{(\phi)} + T^{\mu\nu}_{(\gamma)} \right] = 0, \tag{8.108}$$

which then implies that

$$\left(Q^\nu_{(\phi)} + Q^\nu_{(\gamma)} \right) = 0. \tag{8.109}$$

We have denoted the energy density and pressure associated with the scalar field as ρ_ϕ and p_ϕ. Let ρ_γ and p_γ represent the energy density and pressure associated with radiation. Then, clearly, the total energy density and pressure of the complete system, say, ρ and p will be given by

$$\rho = (\rho_\phi + \rho_\gamma) \quad \text{and} \quad p = (p_\phi + p_\gamma). \tag{8.110}$$

The time component of the conservation equations (8.107) lead to the following equations for the energy density of the scalar field and radiation:

$$\dot\rho_\phi = -3\,H\,(\rho_\phi + p_\phi) + Q_\phi \quad \text{and} \quad \dot\rho_\gamma = -3\,H\,(\rho_\gamma + p_\gamma) + Q_\gamma, \tag{8.111}$$

where Q_ϕ and Q_γ denote the time components of the covariant four vectors $Q^{(\phi)}_\nu$ and $Q^{(\gamma)}_\nu$. Also, due to the constraint (8.109), we have $Q_\phi = -Q_\gamma$, so that

$$\dot\rho = -3\,H\,(\rho + p), \tag{8.112}$$

which, apparently, corresponds to the time component of Eq. (8.108).

Let us now choose $Q_\phi = -(\Gamma \dot{\phi}^2)$, where the quantity Γ is referred to as the decay rate [2, 7]. It effectively describes—albeit, in a classical and in a coarse grained fashion—the perturbative decay of the inflaton into photons and other particles that constitute the standard model which are in thermal equilibrium with radiation. For now, let us assume the decay rate Γ to be a constant[9]. In such a case, the conservation equation (8.111) for ρ_ϕ is given by

$$\dot{\rho}_\phi = -(3H + \Gamma)\,\dot{\phi}^2 = -2\,(3H + \Gamma)\,[\rho_\phi - V(\phi)], \qquad (8.113)$$

and it is useful to note that this equation corresponds to the following modification to the original equation of motion for the scalar field [viz. Eq. (8.8)]:

$$\ddot{\phi} + (3H + \Gamma)\,\dot{\phi} + V_\phi = 0. \qquad (8.114)$$

Let us now return to the situation wherein the scalar field is oscillating at the bottom of the large field inflaton potentials that we had considered in the previous sub-section. When the scalar field is interacting with radiation, upon averaging Eq. (8.113) over the oscillations, we obtain that

$$\dot{\rho}_\phi = -(3H + \Gamma)\,\bar{\alpha}\,\rho_\phi \qquad (8.115)$$

where, as earlier, we have suppressed the angular brackets that denote averaging over the oscillations. This equation for ρ_ϕ can be immediately integrated to arrive at the result

$$\rho_\phi = \rho_0\,(a/a_0)^{-(3\bar{\alpha})}\,\exp - [\Gamma\,\bar{\alpha}\,(t - t_0)], \qquad (8.116)$$

where, recall that, t_0 denotes the time at which the inflaton begins to oscillate. The effect of coupling the scalar field to radiation is evident upon comparing the above expression with the result (8.106) that we had obtained earlier. The coupling leads to an exponential decay of the energy density of the scalar field and, as we shall see below, this energy is transferred to radiation, thereby reheating the universe.

According to the constraint (8.109), the equation (8.113) for ρ_ϕ implies that the energy density of radiation is governed by the equation

$$\dot{\rho}_\gamma = (-4H\,\rho_\gamma + \Gamma\,\bar{\alpha}\,\rho_\phi), \qquad (8.117)$$

---

[9]In the following section, when we consider the evolution of the perturbations during reheating, we shall relax this assumption and assume Γ to be either a constant or dependent on the scalar field.

where we have made use of the fact that $p_\gamma = (\rho_\gamma/3)$. Upon substituting the solution (8.116) for ρ_ϕ, the solution to the above equation for ρ_γ can be expressed as

$$\rho_\gamma(t) = \rho_0 \, (\bar\alpha \, \Gamma) \, (a/a_0)^{-4} \int\limits_{t_0}^{t} d\tilde{t} \, \left[a(\tilde{t})/a_0\right]^{(4-3\,\bar\alpha)} \exp - \left[\bar\alpha \, \Gamma \, (\tilde{t} - t_0)\right],$$

$$(8.118)$$

where we have assumed that $\rho_\gamma(t_0) = 0$. If we now consider a domain wherein the scalar field still dominates the energy density, then we can use the solution (8.106) for ρ_ϕ in the first Friedmann equation to obtain the evolution of the scale factor during this domain to be: $[a(t)/a_0] = (t/t_0)^\beta$, where $\beta = [2/(3\,\bar\alpha)]$. Under this condition, one can carry out the integral in the above expression to obtain (see, for example, the sixth reference in Ref. [2])

$$\rho_\gamma(t) = \rho_0 \, (a/a_0)^{-4} \, (\bar\alpha \, \Gamma \, t_0)^{-(4\beta-2)} \exp(\bar\alpha \, \Gamma \, t_0)$$

$$\times \left(\gamma[(4\beta-1), (\bar\alpha \, \Gamma \, t)] - \gamma[(4\beta-1), (\bar\alpha \, \Gamma \, t_0)]\right), \quad (8.119)$$

where $\gamma(m, x)$ denotes the incomplete Gamma function. For times such that $t_0 < t \ll (\bar\alpha \, \Gamma)^{-1}$, the argument of the incomplete gamma function is small and, upon suitably expanding the gamma function in this limit, one obtains

$$\rho_\gamma(t) = \rho_0 \, [\bar\alpha/(4\beta-1)] \, (\Gamma \, t_0^2/t) \, \left[1 - (t/t_0)^{-(4\beta-1)}\right]. \quad (8.120)$$

As $2 \leq n < \infty$, we have $(5/3) \geq (4\beta-1) > (1/3)$ and, in such a case, it is easy to see that ρ_γ, as given by the above expression, first starts to increase, reaches a maximum and decreases thereafter. Actually, the approximation for the incomplete Gamma function that we have used to arrive at the expression (8.120) ceases to be valid as t approaches $(\bar\alpha \, \Gamma)^{-1}$, since the argument of the Gamma function does not stay small. However, to obtain an order of magnitude estimate, let us stretch the approximation until $t \simeq (\bar\alpha \, \Gamma)^{-1} \equiv t_{\text{RH}}$. In this limit, the second term in the square bracket in Eq. (8.120) proves to be small, and we obtain that

$$\rho_\gamma \simeq \left[\rho_0 \, (\bar\alpha \, \Gamma \, t_0)^2 / (4\beta-1)\right]. \quad (8.121)$$

Once it has attained thermal equilibrium, the energy density of radiation will be given by: $\rho_\gamma^{\text{RD}} = (g_* \, \pi^2 \, T^4/30)$, where g_* denotes the number of relativistic degrees of freedom at the temperature T [1]. As per the first Friedmann equation, we have $\rho_0 \simeq (H_{\text{I}}^2 \, M_{\text{P}}^2)$, where H_{I} is the Hubble scale during inflation. If we now assume that, $t_{\rm o} \simeq H_{\rm I}^{-1}$, then, one obtains the reheating temperature $T_{\rm RH}$ to be [1, 2]

$$T_{\rm RH} \simeq \left(\frac{30}{g_* \pi^2}\right)^{1/4} \left(\frac{\bar{\alpha}^2}{4\,\beta - 1}\right)^{1/4} (\Gamma M_{\rm P})^{1/2}. \qquad (8.122)$$

When radiation has attained such a temperature, the universe can be said to have made the transition to the standard, radiation dominated phase of the hot big bang model. It is worthwhile pointing out that the reheating temperature that we have obtained above does not depend on the energy scale of inflation (viz. $H_{\rm I}$), but depends only on the decay rate Γ of the inflaton. In other words, whatever be the scale at which inflation occurs, given a Γ, the radiation dominated epoch always commences at the same temperature [7, 8].

8.6   Evolution of perturbations during reheating

In this section, we shall consider the evolution of the perturbations during reheating. We shall arrive at the equations describing the large scale curvature perturbations, and explicitly illustrate that perturbative reheating does not affect the amplitude of these perturbations.

Motivated by the discussion in the last section, let us now assume that the function which describes the energy transfer between the inflaton and radiation is given by

$$Q_\phi = -\,(\Gamma\,\rho_\phi) \quad \text{and} \quad Q_\gamma = (\Gamma\,\rho_\phi). \qquad (8.123)$$

Hereafter, let us also assume that the decay rate Γ is either a constant or is dependent on the scalar field. For simplicity, we shall focus on the case wherein $p_\phi = 0$ during the transition from inflation to the radiation dominated epoch. (Such a situation would correspond, for example, to the field oscillating at the bottom of the quadratic potential in the large field models that we had considered in the previous section, which leads to $\alpha = (1/2)$ [cf. Eq. (8.104)] or, equivalently, to a vanishing p_ϕ.) In such a case, the equations that describe the conservation of energy of the two components of the system reduce to

$$\dot{\rho}_\phi = -\,(3\,H + \Gamma)\,\rho_\phi, \quad \text{and} \quad \dot{\rho}_\gamma = (-4\,H\,\rho_\gamma + \Gamma\,\rho_\phi). \qquad (8.124)$$

As it turns out to be more convenient, we shall work with the following dimensionless density parameters [25, 26]

$$\Omega_\phi = (\rho_\phi/\rho) \quad \text{and} \quad \Omega_\gamma = (\rho_\gamma/\rho), \qquad (8.125)$$

and the dimensionless decay rate

$$g \equiv \left(\frac{\Gamma}{\Gamma + H}\right). \tag{8.126}$$

It is useful to note here that $(\Omega_\phi + \Omega_\gamma) = 1$. In terms of the above dimensionless quantities, one finds that the background equations (8.124) can be expressed as follows:

$$\dot{\Omega}_\phi = (H\,\Omega_\phi) \left[\Omega_\gamma - \left(\frac{g}{1-g}\right)\right] \quad \text{and} \quad \dot{\Omega}_\gamma = -(H\,\Omega_\phi)\left[\Omega_\gamma - \left(\frac{g}{1-g}\right)\right], \tag{8.127}$$

whereas g is found to satisfy the equation

$$\dot{g} = \left(\frac{H}{2}\right)(4 - \Omega_\phi)\,g\,(1-g) + \left(\frac{\dot{\Gamma}}{\Gamma}\right)g\,(1-g). \tag{8.128}$$

8.6.1 *Equations governing the evolution of perturbations*

Let us now turn to the derivation of the equations governing the evolution of the scalar perturbations post-inflation. As we had pointed out, current observations seem to indicate that the largest scale today would have left the Hubble radius not more than 60 or so e-folds before the end of inflation. It is straightforward to establish that, during inflation, all the cosmologically relevant scales leave the Hubble radius over an interval of about 8-10 e-folds. So, if these scales leave the Hubble radius during the early stages of the inflationary epoch, they are well outside the Hubble radius during reheating.

Earlier, during inflation, we had chosen the curvature perturbation $\mathcal{R}$ [cf. Eq. (8.40)] to describe the scalar perturbations. On super-Hubble scales, it turns out to be more convenient to consider the following gauge invariant quantity ζ to describe the scalar perturbations:

$$\zeta = -\Psi - H\left(\frac{\delta\varrho}{\dot{\rho}}\right), \tag{8.129}$$

which denotes the curvature perturbation on uniform total density hypersurfaces [27]. In the situation of our interest, $\delta\varrho = (\delta\varrho_\phi + \delta\varrho_\gamma)$. In the absence of anisotropic stresses, upon using the first order Einstein's equations (8.36), one can readily show that [1, 2]

$$\zeta = -\mathcal{R} - \left(\frac{1}{4\pi G\,a^2}\right)\left(\frac{H}{\dot{\rho}}\right)\nabla^2\Phi. \tag{8.130}$$

In other words, on super-Hubble scales [i.e. for $(k/a\,H) \ll 1$], $\zeta \simeq -\mathcal{R}$ to a very good approximation. The curvature perturbation associated with the scalar field and radiation can be defined as

$$\zeta_\phi = -\Psi - H\left(\frac{\delta\varrho_\phi}{\dot\rho_\phi}\right) \quad \text{and} \quad \zeta_\gamma = -\Psi - H\left(\frac{\delta\varrho_\gamma}{\dot\rho_\gamma}\right), \qquad (8.131)$$

so that the total curvature perturbation ζ is a weighted sum of the two as follows:

$$\zeta = \left(\frac{\dot\rho_\phi}{\dot\rho}\right)\zeta_\phi + \left(\frac{\dot\rho_\gamma}{\dot\rho}\right)\zeta_\gamma. \qquad (8.132)$$

In the line-element (8.28) that describes the Friedmann universe in the presence of the scalar perturbations, the conservation equations (8.107) lead to the following equations for the gauge invariant perturbations in the energy densities of the scalar field and radiation:

$$\dot{\delta\varrho}_\phi + 3\,H\,(\delta\varrho_\phi + \delta\mathcal{P}_\phi) - \left(\frac{1}{a^2}\right)\nabla^2\delta\varsigma_\phi$$
$$- 3\,(\rho_\phi + p_\phi)\,\dot\Psi = (\delta\mathcal{Q}_\phi + Q_\phi\,\Phi), \qquad (8.133a)$$

$$\dot{\delta\varrho}_\gamma + 3\,H\,(\delta\varrho_\gamma + \delta\mathcal{P}_\gamma) - \left(\frac{1}{a^2}\right)\nabla^2\delta\varsigma_\gamma$$
$$- 3\,(\rho_\gamma + p_\gamma)\,\dot\Psi = (\delta\mathcal{Q}_\gamma + Q_\gamma\,\Phi), \qquad (8.133b)$$

where $\delta\mathcal{Q}_\phi$ and $\delta\mathcal{Q}_\gamma$ denote the gauge invariant first order perturbations in the quantities Q_ϕ and Q_γ that represent the transfer of energy between the two components. It can be shown that, under the scalar gauge transformations (8.29), the perturbations δQ_ϕ and δQ_γ transform as follows:

$$\delta Q_\phi \to \left(\delta Q_\phi - \dot Q_\phi\,\delta t\right) \quad \text{and} \quad \delta Q_\gamma \to \left(\delta Q_\gamma - \dot Q_\gamma\,\delta t\right) \qquad (8.134)$$

so that the corresponding gauge invariant perturbations $\delta\mathcal{Q}_\phi$ and $\delta\mathcal{Q}_\gamma$ are found to be

$$\delta\mathcal{Q}_\phi = \delta Q_\phi + \dot Q_\phi\left[a\,(B - a\,\dot E)\right] \quad \text{and} \quad \delta\mathcal{Q}_\gamma = \delta Q_\gamma + \dot Q_\phi\left[a\,(B - a\,\dot E)\right]. \qquad (8.135)$$

Recall that, we are focusing on the situation wherein p_ϕ is zero. In such a case, the corresponding perturbation in the pressure too vanishes identically. Under these conditions, on large scales [i.e. when $(k/a\,H) \ll 1$], upon using the standard equation of state for radiation, the above equations (8.133) reduce to

$$\dot{\delta\varrho}_\phi + 3\,H\,\delta\varrho_\phi - 3\,\rho_\phi\,\dot\Psi = (\delta\mathcal{Q}_\phi + Q_\phi\,\Phi), \qquad (8.136a)$$

$$\dot{\delta\varrho}_\gamma + 4\,H\,\delta\varrho_\gamma - 4\,\rho_\gamma\,\dot\Psi = (\delta\mathcal{Q}_\gamma + Q_\gamma\,\Phi). \qquad (8.136b)$$

I should mention that, since the inflaton and radiation have fixed equations of state, they do not possess any intrinsic non-adiabatic pressure perturbations. Upon using evolution equations (8.136) and the expressions (8.123) for Q_ϕ and Q_γ, one can arrive at the equations governing the curvature perturbations on uniform inflaton and radiation density hypersurfaces, viz. ζ_ϕ and ζ_γ, respectively. They can be written as [25, 26]

$$\dot{\zeta}_\phi = -\left(\frac{\Gamma \dot{\rho}_\gamma}{6 \dot{\rho}_\phi}\right)\left(\frac{\rho_\phi}{\rho}\right) S_{\phi\gamma} + \left(\frac{H \rho_\phi}{\dot{\rho}_\phi}\right) \delta \Upsilon_\phi, \qquad (8.137\text{a})$$

$$\dot{\zeta}_\gamma = \left(\frac{\Gamma \dot{\rho}_\phi}{3 \dot{\rho}_\gamma}\right)\left[1 - \left(\frac{\rho_\phi}{2\rho}\right)\right] S_{\phi\gamma} - \left(\frac{H \rho_\phi}{\dot{\rho}_\gamma}\right) \delta \Upsilon_\gamma, \qquad (8.137\text{b})$$

where the quantity $S_{\phi\gamma}$ represents the gauge invariant relative entropy (i.e. the iso-curvature) perturbation between the scalar field and radiation, and is given by

$$S_{\phi\gamma} = 3\,(\zeta_\phi - \zeta_\gamma) = -3\,H\left[\left(\frac{\delta \varrho_\phi}{\dot{\rho}_\phi}\right) - \left(\frac{\delta \varrho_\gamma}{\dot{\rho}_\gamma}\right)\right]. \qquad (8.138)$$

Moreover, the quantities $\delta \Upsilon_\phi$ and $\delta \Upsilon_\gamma$ are defined as

$$\delta \Upsilon_\phi = \delta \Upsilon - \dot{\Gamma}\left(\frac{\delta \varrho_\phi}{\dot{\rho}_\phi}\right) \quad \text{and} \quad \delta \Upsilon_\gamma = \delta \Upsilon - \dot{\Gamma}\left(\frac{\delta \varrho_\gamma}{\dot{\rho}_\gamma}\right), \qquad (8.139)$$

with

$$\delta \Upsilon = \delta \Gamma + \dot{\Gamma}\left[a\,(B - a\,\dot{E})\right] \qquad (8.140)$$

being the gauge invariant perturbation in the decay rate of the inflaton.

From the definition of the total curvature perturbation (8.132), the iso-curvature perturbation (8.138), and the equations (8.127), (8.128) and (8.137) that describe the background and the perturbations, one can then obtain the following system of equations governing the curvature perturbations ζ and ζ_ϕ [25, 26]:

$$\dot{\zeta} = \left[-\left(\frac{(3 - 2\,g)\,H\,\Omega_\phi}{(1 - g)\,(4 - \Omega_\phi)}\right)\right.$$
$$\left. + \left(\frac{\dot{\Gamma}\,(1 - g)}{(3 - 2\,g)\,H}\right)\left(\frac{4\,(1 - g) + (4 - 3\,g)\,\Omega_\phi}{4\,(1 - g)\,(1 - \Omega_\phi) - g\,\Omega_\phi}\right)\right] (\zeta - \zeta_\phi), \qquad (8.141\text{a})$$

$$\dot{\zeta}_\phi = \left(\frac{g\,H\,(4 - \Omega_\phi)}{2\,(3 - 2\,g)}\right)(\zeta - \zeta_\phi) - \left(\frac{1 - g}{3 - 2\,g}\right) \delta \Upsilon_\phi. \qquad (8.141\text{b})$$

Using these equations, let us now turn to understanding the effects of reheating on the evolution of the perturbations.

8.6.2   *Effects of reheating on the perturbations*

At the beginning of the epoch of reheating, both the background energy density and the perturbations are dominated by the contribution due to the scalar field so that, at this stage, we have $\rho \simeq \rho_\phi$ and $\zeta \simeq \zeta_\phi$. The quantity ζ_ϕ is essentially determined by the curvature perturbation due to the scalar field, say, $\mathcal{R}_I$, evaluated when the modes are well outside the Hubble radius during inflation. As reheating proceeds and, as the scalar field is gradually being converted into radiation, the perturbations in the scalar field are also transferred to the perturbations in the radiation. An important issue that needs to be understood is whether the process of reheating that we had discussed in the previous section modifies the amplitude of the large scale curvature perturbations. Let us now consider the effects of reheating due to the two possible cases, viz. when the decay rate Γ is either a constant or is a function of the inflaton.

When Γ is a constant, $\dot{\Gamma}$ and $\delta\Gamma$ are zero and, hence, $\delta\Upsilon_\phi$ vanishes identically [cf. Eq. (8.139)]. In such a situation, it is clear from the equations (8.141) that $\zeta \simeq \zeta_\phi$ is an attractor. In other words, for scales of cosmological interest, the total curvature perturbation ζ remains constant even during reheating. (It is worth recalling here that the curvature perturbation $\mathcal{R}$ is conserved on super-Hubble scales during inflation.) Therefore, while $\zeta_\phi \simeq -\mathcal{R}_I$ during the early phase of reheating, by the start of the radiation dominated regime, the scalar field has all but disappeared and, hence, during this stage $\rho \simeq \rho_\gamma$, while $\zeta \simeq \zeta_\gamma \simeq -\mathcal{R}_I$.

On the other hand, if the decay rate Γ depends on the scalar field, then the gauge invariant perturbation $\delta\Upsilon_\phi$ can be written as

$$\delta\Upsilon_\phi = \left(\frac{\partial\Gamma}{\partial\phi}\right)\dot{\phi}\left(\frac{\delta\varphi}{\dot{\phi}} - \frac{\delta\varrho_\phi}{\dot{\rho}_\phi}\right), \qquad (8.142)$$

where $\delta\varphi$ is the gauge invariant perturbation in the scalar field [cf. Eq. (8.66)]. It is straightforward to establish that, under the conditions we are working in, viz. $p_\phi = 0$ and $\delta\mathcal{P}_\phi = 0$, $(\delta\varrho_\phi/\dot{\rho}_\phi) = (\delta\varphi/\dot{\phi})$, so that $\delta\Upsilon_\phi$ vanishes identically. Hence, $\zeta \simeq \zeta_\phi$ is an attractor just as in the case where Γ was a constant. Therefore, in both the possible situations of our interest, i.e. when the decay rate is either a constant or dependent on the scalar field, the total, large scale curvature perturbation ζ remains conserved during the epoch of reheating. So, in such scenarios, at the early stages of the radiation dominated epoch, ζ is basically determined by the curvature perturbation $\mathcal{R}_I$ generated during inflation.

8.7   Non-trivial post-inflationary dynamics: Modulated reheating and the curvaton scenarios

In the previous section, we had illustrated as to how the simple mechanism of reheating does not affect the amplitude of the large scale perturbations. In this section, we shall discuss two scenarios—viz. the modulated reheating and the curvaton scenarios—that can affect the amplitude of the perturbations post inflation. As we shall see, while the modulated reheating mechanism induces new perturbations during reheating, the curvaton mechanism converts initial iso-curvature perturbations into curvature perturbations.

8.7.1   *Modulated or the inhomogeneous reheating scenario*

Consider the case wherein the decay rate Γ depends on a scalar field other than the inflaton so that, while $\dot{\Gamma} = 0$, $\delta\Upsilon_\phi = \delta\Upsilon$. Such a possibility can arise if the inflaton decay rate depends on the vacuum expectation value of another field whose quantum fluctuations are excited during inflation, but whose time variation during reheating is negligible [26, 28]. When the inflaton is oscillating at the minimum of its potential during the early stages of reheating, we can set $\Omega_\phi \simeq 1$ and $g \simeq 0$. In such a case, the equations (8.141) reduce to

$$\dot{\zeta} \simeq -H \left(\zeta - \zeta_\phi\right) \quad \text{and} \quad \dot{\zeta}_\phi \simeq -\left(\frac{\delta\Upsilon}{3}\right). \tag{8.143}$$

Note that, on super-Hubble scales, one can assume that $\delta\Upsilon$ is almost a constant. Further, since we are confining our attention to the case wherein the inflaton is described by a quadratic potential, we have $a \propto t^{2/3}$ when the scalar field is oscillating at the bottom of the potential. The above equations can be easily integrated under these conditions to arrive at the result that, when $t \simeq \Gamma^{-1}$,

$$\zeta \simeq \zeta_i - \left(\frac{2}{15}\right)\left(\frac{\delta\Upsilon}{\Gamma}\right), \tag{8.144}$$

where ζ_i denotes the initial curvature perturbation, say, when the inflaton begins to oscillate. Recall that, during the radiation dominated epoch, the Bardeen potential is given in terms of the curvature perturbation by the relation [cf. Eq. (8.54)]: $\Phi^R \simeq (2/3)\,\mathcal{R} \simeq -(2/3)\,\zeta$, with the last expression arising due to the fact that $\zeta \simeq -\mathcal{R}$ on the scales of our interest. If we now further assume that ζ_i is small, then, from the solution (8.144), we obtain that

$$\Phi^{\mathrm{R}} \simeq \left(\frac{4}{45}\right)\left(\frac{\delta\Upsilon}{\Gamma}\right) \simeq \left(\frac{1}{9}\right)\left(\frac{\delta\Upsilon}{\Gamma}\right). \qquad (8.145)$$

In other words, perturbations can be generated due to the modulations or the inhomogeneities in the decay rate during the reheating phase (and, hence, the name to the mechanism) [26, 28]. The perturbations thus generated can supplement the inflationary perturbations that may already be present.

8.7.2   *The curvaton scenario*

The curvaton scenario essentially involves a second scalar field (i.e. apart from the inflaton) that converts an initial iso-curvature perturbation into curvature perturbation (and, hence, such a scalar field is referred to as the curvaton). Consider a scalar field ϕ (which is *not* the inflaton) that is oscillating at the bottom of its potential and is interacting with radiation through a constant decay rate Γ during the early stages of the radiation dominated epoch. In such a case, while Ω_ϕ and Ω_γ continue to be governed by Eqs. (8.127), the equation (8.128) describing the evolution of the quantity g reduces to

$$\dot{g} = \left(\frac{H}{2}\right)(4 - \Omega_\phi)\, g\, (1 - g). \qquad (8.146)$$

Also, when Γ is a constant, Eqs. (8.141) that govern the evolution of the perturbations simplify to

$$\dot{\zeta} = -\left(\frac{(3 - 2g)\, H\, \Omega_\phi}{(1 - g)\, (4 - \Omega_\phi)}\right)(\zeta - \zeta_\phi), \qquad (8.147\mathrm{a})$$

$$\dot{\zeta}_\phi = \left(\frac{g\, H\, (4 - \Omega_\phi)}{2\, (3 - 2g)}\right)(\zeta - \zeta_\phi). \qquad (8.147\mathrm{b})$$

An analysis of the background equations indicates that they admit three fixed points corresponding to the following sets of values of $(\Omega_\phi, \Omega_\gamma, g)$: $(0, 1, 0)$, $(1, 0, 0)$ and $(0, 1, 1)$ [25]. For convenience, let us refer to these three points as Ⓐ, Ⓑ and Ⓒ, respectively. Evidently, while the domains near Ⓐ and Ⓒ correspond to regimes of radiation domination, the domains close to Ⓑ correspond to epochs where the scalar field is dominant. It can be shown that, while the fixed point Ⓐ is an unstable repellor, Ⓒ is a stable attractor, whereas Ⓑ is a saddle point. A generic solution starts at Ⓐ and approaches Ⓒ at late times. The conventional radiation dominated epoch, of course, corresponds to $\Omega_\phi = 0$. However, in the presence of the scalar field, it is found that a solution starting at Ⓐ can approach the scalar field dominated domain Ⓑ arbitrarily closely, before the scalar field decays, thereby ending at Ⓒ.

Consider a background trajectory that starts close to Ⓐ. Let Ω_ϕ^i be the initial value of the dimensionless density parameter associated with the scalar field. Also, let us assume that the scalar field is initially perturbed, while radiation is not, so that we have $\zeta_\gamma = 0$ and $\zeta_\phi = \zeta_i$, say. In such a case, we have

$$\zeta = \left(\frac{3\,\Omega_\phi^i}{4 - \Omega_\phi^i} \right) \zeta_i \quad \text{and} \quad \mathcal{S}_{\phi\gamma} = 3\,\zeta_i, \tag{8.148}$$

which corresponds to an initial iso-curvature perturbation since $\zeta \to 0$ at early times as Ω_ϕ^i is small in this domain. Under these conditions, upon evolving Eqs. (8.147) for ζ and ζ_ϕ, at late times, as the background approaches the attractor Ⓒ, one obtains that, $\zeta = (\mathcal{D}\,\zeta_i)$, while $\mathcal{S}_{\phi\gamma} = 0$, where the constant of proportionality $\mathcal{D}$, though it depends on the details of the background dynamics, can be of order unity [25, 29]. Such a perturbation would correspond to an adiabatic perturbation. Clearly, the presence of the scalar field allows for the conversion of an initial iso-curvature perturbation into a purely adiabatic, curvature perturbation.

8.8   Summary and discussion

Let me now conclude with a brief summary, and also highlight the motivations behind the recent upsurge in interest in studying perturbation theory beyond the linear order.

In this article, I have presented an overview of the origin and evolution of perturbations during inflation and reheating. I have outlined as to how a suitable duration of inflation can resolve the horizon problem that arises in the standard hot big bang model, and have discussed the generation of a nearly scale invariant spectrum of the scalar and the tensor perturbations in the scalar field driven, slow roll, inflationary scenario. I have also illustrated that, while perturbative reheating does not affect the large scale curvature perturbations, there exist certain non-trivial, post-inflationary, dynamics that can lead to a change in the amplitude of the perturbations.

As I have repeatedly emphasized, in this article, I have confined my discussion to perturbation theory at the linear order. Moreover, as I have pointed out, at the linear order, perturbations remain Gaussian. However, analysis of the recent CMBR data seem to indicate that deviations from Gaussianity may possibly be large (see, for instance, Refs. [5, 30]). Further, it has been realized that, if future observations indeed confirm such a large level of non-Gaussianity, then, it can result in a substantial tightening in the constraints on the inflationary models as well as the post-inflationary scenarios. Understanding the extent of non-Gaussianity generated during the inflationary and the post-inflationary regimes requires a systematic study of the perturbations at the higher orders. It is for this reason that, over the last few years, a considerable amount of effort has been devoted to the study of perturbation theory beyond the linear order and their imprints on the CMBR (see, for example, Refs. [19, 30]).

Acknowledgements

I would like to take this opportunity to thank T. Padmanabhan who has had a deep influence in shaping my outlook, not only on physics, but also in my personal life. I also wish to acknowledge collaborations or discussions with Raul Abramo, Moumita Aich, Pravabati Chingangbam, Jinn-Ouk Gong, Dhiraj Hazra, Rajeev Jain, Will Kinney, Marc Lilley, Jérôme Martin, Patrick Peter, Tarun Souradeep, S. Shankaranarayanan, Kandaswamy Subramanian and Sanil Unnikrishnan, on related issues at different stages.

References

[1] E. W. Kolb and M. S. Turner, *The Early Universe* (Addison-Wesley, Redwood City, California, 1990); S. Dodelson, *Modern Cosmology* (Academic Press, San Diego, U.S.A., 2003); V. F. Mukhanov, *Physical Foundations of Cosmology* (Cambridge University Press, Cambridge, England, 2005); S. Weinberg, *Cosmology* (Oxford University Press, Oxford, England, 2008); R. Durrer, *The Cosmic Microwave Background* (Cambridge University Press, Cambridge, England, 2008); D. H. Lyth and A. R. Liddle, *The Primordial Density Perturbation* (Cambridge University Press, Cambridge, England, 2009); P. Peter, J-P. Uzan and J. Brujic, *Primordial Cosmology* (Oxford University Press, Oxford, England, 2009).

[2] H. Kodama and M. Sasaki, Prog. Theor. Phys. Suppl. **78**, 1 (1984); V. F. Mukhanov, H. A. Feldman and R. H. Brandenberger, Phys. Rep. **215**, 203 (1992); J. E. Lidsey, A. Liddle, E. W. Kolb, E. J. Copeland,

T. Barreiro and M. Abney, Rev. Mod. Phys. **69**, 373 (1997); D. H. Lyth and A. Riotto, Phys. Rep. **314**, 1 (1999); A. Riotto, arXiv:hep-ph/0210162; J. Martin, arXiv:hep-th/0406011; B. Bassett, S. Tsujikawa and D. Wands, Rev. Mod. Phys. **78**, 537 (2006); W. H. Kinney, arXiv:0902.1529 [astro-ph.CO]; L. Sriramkumar, Curr. Sci. **97**, 868 (2009); D. Baumann, arXiv:0907.5424v1 [hep-th].

[3] A. A. Starobinsky, JETP Lett. **30**, 682 (1979); Phys. Lett. B **91**, 99 (1980); D. Kazanas, Astrophys. J. **241**, L59 (1980); K. Sato, Mon. Not. Roy. Astron. Soc. **195**, 467 (1981); A. Guth, Phys. Rev. D **23**, 347 (1981); A. D. Linde, Phys. Lett. B **108**, 389 (1982); *ibid.* **114**, 431 (1982); Phys. Rev. Lett. **48**, 335 (1982); A. Albrecht and P. Steinhardt, Phys. Rev. Lett. **48**, 1220 (1982).

[4] V. F. Mukhanov and G. V. Chibisov, Sov. Phys. JETP Lett. **33**, 532 (1981); S. W. Hawking, Phys. Lett. B **115**, 295 (1982); A. A. Starobinsky, Phys. Lett. B **117**, 175 (1982); A. Guth and S.-Y. Pi, Phys. Rev. Lett. **49**, 1110 (1982).

[5] D. Larson *et al.*, Astrophys. J. Suppl. **192**, 16 (2011); E. Komatsu *et al.*, Astrophys. J. Suppl. **192**, 18 (2011).

[6] I. G. Moss, Phys. Lett. B, **154**, 120 (1985); A. Berera and L. Z. Fang, Phys. Rev. Lett. **74**, 1912 (1985); L. M. H. Hall, I. G. Moss and A. Berera, Phys. Rev. D **69**, 083525 (2004).

[7] M. S. Turner, Phys. Rev. D **28**, 1243 (1983); A. Albrecht, P. J. Steinhardt, M. S. Turner and F. Wilczek, Phys. Rev. Lett. **48**, 1437 (1982).

[8] J. H. Traschen and R. H. Brandenberger, Phys. Rev. D **42**, 2491 (1990); L. Kofman, A. Linde and A. Starobinsky, Phys. Rev. D **56**, 3258 (1997).

[9] S. Dodelson and L. Hui, Phys. Rev. Letts. **91**, 131301 (2003); A. R. Liddle and S. M. Leach, Phys. Rev. D **68**, 103503 (2003).

[10] P. J. Steinhardt and M. S. Turner, Phys. Rev. D **29**, 2162 (1984); D. S. Salopek and J. R. Bond, Phys. Rev. D **42**, 3936 (1990); A. R. Liddle and D. H. Lyth, Phys. Letts. B **291**, 391 (1992); A. R. Liddle, P. Parsons and J. D. Barrow, Phys. Rev. D **50**, 7222 (1994); D. J. Schwarz, C. A. Terrero-Escalante and A. A. Garcia, Phys. Lett. B **517**, 243 (2001); S. M. Leach, A. R. Liddle, J. Martin and D. J. Schwarz, Phys. Rev. D **66**, 023515 (2002).

[11] A. Linde, Phys. Letts. B **129**, 177 (1983).

[12] J. Bardeen, Phys. Rev. D **22**, 1882 (1980).

[13] J. M. Stewart, Class. Quantum Grav. **7**, 1169 (1990).

[14] L. P. Grishchuk, Sov. Phys. JETP **40**, 409 (1974); A. A. Starobinsky, Sov. Phys. JETP Lett. **30**, 682 (1979); V. A. Rubakov, M. V. Sazhin

and A. V. Veryaskin, Phys. Lett. B **115**, 189 (1982).

[15] C. Gordon, D. Wands, B. A. Bassett and R. Maartens, Phys. Rev. D **63**, 023506 (2001).

[16] S. Unnikrishnan and L. Sriramkumar, Phys. Rev. D **81**, 103511 (2010).

[17] V. N. Lukash, Sov. Phys. JETP **52**, 807 (1980); D. H. Lyth, Phys. Rev. D **31**, 1792 (1985).

[18] V. S. Mukhanov, JETP Lett. **41**, 493 (1985); M. Sasaki, Prog. Theor. Phys. **76**, 1036 (1986).

[19] N. Bartolo, S. Matarrese and A. Riotto, arXiv:1001.3957v1 [astro-ph.CO]; N. Bartolo, E. Komatsu, S. Matarrese and A. Riotto, Phys. Rep. **402**, 103 (2004); J. Maldacena, JHEP **0305**, 013 (2003); D. Seery and J. E. Lidsey, JCAP **0506**, 003 (2005); X. Chen, Phys. Rev. D **72**, 123518 (2005); X. Chen, M.-x. Huang, S. Kachru and G. Shiu, JCAP **0701**, 002 (2007); D. Langlois, S. Renaux-Petel, D. A. Steer and T. Tanaka, Phys. Rev. Lett. **101**, 061301 (2008); Phys. Rev. D **78**, 063523 (2008); X. Chen, Adv. Astron. **2010**, 638979 (2010);

[20] S. M. Leach and A. R. Liddle, Phys. Rev. D **63**, 043508 (2001); S. M. Leach, M. Sasaki, D. Wands and A. R. Liddle, Phys. Rev. D **64**, 023512 (2001); R. K. Jain, P. Chingangbam and L. Sriramkumar, JCAP **0710**, 003 (2007); A. J. Christopherson and K. A. Malik, Phys. Lett. B **675**, 159 (2009).

[21] E. D. Stewart and D. H. Lyth, Phys. Letts. B **302**, 171 (1993); J. C. Hwang and H. Noh, Phys. Rev. D **54**, 1460 (1996).

[22] T. Bunch and P. C. W. Davies, Proc. Roy. Soc. Lond. A **360**, 117 (1978).

[23] L. F. Abbott and M. B. Wise, Nucl. Phys. B **244**, 541 (1984); D. H. Lyth and E. D. Stewart, Phys. Lett. B **274**, 168 (1992); J. Martin and D. J. Schwarz, Phys. Rev. D **57**, 3302 (1998); L. Sriramkumar and T. Padmanabhan, Phys. Rev. D **71**, 103512 (2005).

[24] E. F. Bunn, A. R. Liddle and M. J. White, Phys. Rev. D **54**, R5917 (1996).

[25] K. A. Malik, D. Wands and C. Ungarelli, Phys. Rev. D **67**, 063516 (2003).

[26] S. Matarrese and A. Riotto, JCAP **08**, 007 (2003).

[27] J. M. Bardeen, P. J. Steinhardt and M. S. Turner, Phys. Rev. D **28**, 679 (1983).

[28] A. Mazumdar and M. Postma, Phys. Lett. B **573**, 12 (2003); G. Dvali, A. Gruzinov and M. Zaldarriaga, Phys. Rev. D **69**, 023505 (2004).

[29] D. H. Lyth and D. Wands, Phys. Lett. B **524**, 5 (2002); D. H. Lyth,

C. Ungarelli and D. Wands, Phys. Rev. D **67**, 023503 (2003).

[30] A. P. S. Yadav and B. D. Wandelt, Phys. Rev. Lett. **100**, 181301 (2008); K. M. Smith, L. Senatore and M. Zaldarriaga, JCAP **0909**, 006 (2009); M. Liguori, E. Sefusatti, J. R. Fergusson and E. P. S. Shellard, Adv. Astron. **2010**, 980523 (2010); A. P. S. Yadav and B. D. Wandelt, arXiv:1006.0275v3 [astro-ph.CO]; E. Komatsu, Class. Quantum Grav. **27**, 124010 (2010).

Chapter 9

Patterns in neural processing

Sunu Engineer

CEO, Embedded Computing Machines Pvt. Ltd.,
19/588 New Era Society, Pune 411 037, Maharashtra, India.
E-mail: *sunu@embeddedmachines.com*

Abstract: In this paper we propose a model for neural processing that addresses both the evolutionary and functional aspects of neural systems that are observed in nature, from the simplest neural collections to dense large scale associations such as human brains. We propose both an architecture and a process in which these components interact to create the emergent behavior that we define as the 'mind'.

9.1 Introduction

Although, over the many years that one has observed the existence of a remarkable set of cells interacting together to form the brain, a clear model for how they function collectively has not emerged. We are familiar with the extant results describing their microscopic functionality in terms of how they assimilate input signals and create output signals, both of which are characterized by excess electromagnetic charge in terms of ions, thus enabling the classification of the brain as an electromagnetic device, made up of many similar but not equal interacting parts.

As extensive research has elucidated the nature of the brain itself, the question of its operations has resolutely evaded a complete characterization. What is important to note is that natural language capability which we consider as a prime characterization of intelligence or higher order brain function [1] is still unrealizable in a digital computation framework inspite of the enormous increases in speed that the underlying technology has achieved. The other important observation is the sheer ubiquity of the neural collections that we call 'brains' in the world, and the large variety of input output behavior that they manifest, from simple reaction behavior to complex symbolic translation, manipulation,interpolation and extrapolation functions. The ubiquity and the wide variety of behavior begs an explanation in terms of a more fundamental unifying theory describing brain function in all such assemblies. In addition we have to elucidate how such complex manipulative structures could be accommodated in an evolutionary framework. This is the basic attempt of this paper.

We expect that the architecture of these neural complexes described below, will enable us to clearly define 'mind' and at a later stage connect the 'mind' to the molecular processes mediated by genes and other biochemical structures. This should permit us to describe and derive the wide variety of behavior that is observed in nature. This includes the higher order thinking in terms of problem solving and recognizing structural similarity between problems, and developing common techniques or variations to solve problems.

In simpler terms, we know how the pieces work but we are not really sure what can come out when we put them all together and they get going. We have worked at understanding this 'device' in terms of its observed functionality as well, and there are huge tomes cataloguing it. One of the questions that we should like to ask is this, is this catalog complete? Have we discovered all the different things that this assembly of 'neurons' can do?

Another way of looking at this problem is to ask, can we specify the complete functionality of the 'brain' in terms of its responses to inputs? It is quite obvious from our current levels of knowledge that this is not possible even for simple aggregates of neurons.

9.2   The brain—the neuron

In terms of evolution and in terms of analysis we require a simple and universal model for the brain. The important thing to consider here is the way in which such a system can be created or evolved rather than finding a way to describe every known phenomena using this model. The descriptive approach, that has been the conventional epistemology, creates a huge body of data, where selection of the relevancy becomes a problem in its own right.

In terms of formation and development of the brain there is considerable work in dealing with sizes, the parameter space characterizing them, the correlations between the parameters, the heterochronic aspects of the evolution and so on. However, given this we need to be able to explain this data from a basic unifying model.

We have to consider both the synchronic and diachronic aspect of the brain as a device. What we have today in terms of the neural complex is as follows: all these complexes are made of 'neurons' which are cellular in nature and are capable of electromagnetic interactions with each other via charge transfer (neurotransmitters) across specialized parts of the cell membrane, some inward and some outward. The outward ports of these neurons are connected to the inward ports of others and these asymmetric connections between them form an asymmetric interconnected structure.

The size and number of neurons and brain sizes are correlated in many studies with a variety of parameters such as body size, olfactory cortex size, the neurogenesis pathways and so on. Here we have the interesting speculation that needs to be worked out—the complexity of the advanced brain. Is it only a function of the brain size and number of connections or do the connections themselves change their nature and consequently the nature of the processing itself changes? In addition the changed protein structure at the synapses may be correlated with changed modes of processing via quantum entanglement processes akin to those proposed in [2].

Let us take a simpler model to construct 'brains' in which we ignore the complex aspects of organization of these neural complexes into subsystems and shall focus on the basic physical description of these complexes and not on the functional substructure as observed in various experiments. Our attempts will be to propose a model in which this functional localization (and the corresponding observations about the sizes of these 'subsystems') can be obtained as an emergent phenomenon.

One possible mathematical characterization of these systems can be effected in terms of graphs. At the basic level, the graph consists of the nodes (the nodes are characterized by the number and types of connections, input and output) and the interconnections between them which are classically modeled as weights attached to the interconnections. However a description of the dynamics of the vertices and edges of these graphs are too cumbersome and does not lead to any clear insight into the behavior of the system as a whole. We can categorize these 'brains' in terms of simpler parameter spaces such as the number of neurons, number of interconnects etc. However, lacking a characterization of the emergent functionality, it becomes difficult to establish correlations between the parameter space of the underlying graph and the behavior of the system as a whole. What we need to do is to characterize the functional behavior of the 'brain' itself and then correlate it with the microscopic dynamics of the neurons, in a manner similar to the connection between thermodynamics and statistical mechanics in physical theories of the universe. A model for such a characterization is proposed in this paper.

To try and establish the evolutionary model underlying these 'brains' let us begin with the assumption that just like the sensory organs and their myriad variations evolved in response to the physical and chemical universe the brain evolved in response to something. The question that we want answered is what did the brain evolve in response to? And what form did this response take?

Our thesis here is **that the brain evolved in response to what we call spatio-temporal patterns immanent in the random space of events**. These spatio temporal patterns exist in the universe in the sense of higher order correlations between events in space time and the brain evolved in order to take advantage of these patterns. If we look at the set of spacetime events and compute the set of correlations between them (defined in a general continuous sense), we obtain a set of non-trivial correlations that we refer to as patterns. These patterns are what the brain is designed to operate on.

Since the space time events are mirrored or translated into the brain space by the senses, the patterns also manifest as correlations between different senses and create correlations between the real (as recorded by sensing systems) and the imaginary ones (not recorded by the sensing systems yet present as integral part of brain processing). For example, this would explain synethesia very well.

We hypothesize that initially the correlations and the patterns were manifest as sympathetic activation of different sensory systems (which were all weakly coupled together) and then the system evolved further to effect a separation between the system which computes the correlation, 'the brain' and the sensory organs, 'the senses' [3]. What the above implies is that for lower complexity neural complexes around us, we will find no separate higher order correlation functionality, rendering them 'reactive', as opposed to a 'brain' which computes correlations after an appropriate symbolic translation which we shall refer to as 'thinking'. What parameter values determine the transition? What is the nature of this transition? These are all questions whose answers will occupy us going forward.

If the system evolved to process patterns, or spatio temporal correlations and was complex enough to be able to compute higher order correlations between the patterns themselves, for instance by a coarse averaging model (a renormalization kind of approach will work very well here) and computing the correlations between the averaged field, it could easily explain the evolution of music and language which are fairly easily modeled as higher order, above that of undivided sensations which are auditory and visual. This 'averaging' probably defines a methodology for creating higher order processes, like language and even larger scale and higher order correlations like many of our emotional states etc, over it. However the point to be borne in mind is the following: music, language and other 'intelligent' processes are characterized by what we could refer to as a level of 'translation'. Remapping of the patterns into other isomorphisms (weak isomorphisms or approximations as well) and processing the isomorphisms and remapping them into the original pattern space. This is probably the key to evolution of 'linguistic intelligence'. How this is correlated with the properties of the 'brain' is a problem whose solution is still elusive.

In order to proceed further,if patterns are the 'atoms' or basic building blocks of neural activity we need to define the rules that these patterns follow, namely their algebra and calculus. We must in addition define the dynamical equations obeyed by these patterns which determines the time evolution of these patterns. An interesting question here is the notion of time itself. We should be able to derive time perception in terms of the relationships between patterns [4].

One critical difference between our normal forms of algebra and calculus and this pattern based computation is the fact the whole of the 'atom' is not present as the computation is executed. What we hypothesize as happening is effectively this: the computation proceeds along with the patterns being presented to the engine in a continuous fashion (even though the system is robust to variations in the speed with which the pattern is presented there probably is a low speed threshold below which the notion of the pattern might be difficult to sustain just as the notion of motion using still photoframes is difficult to sustain at a low rate of frames per second).

So we require a calculus or algebra for these patterns. One in which the operation is time dependent and the result evolves in time continuously as more of the pattern is presented to the input system.

A potential way of modeling it is as a dynamical system which is evolving under time dependent and changing boundary conditions.

For example when memory is modeled in terms of patterns as an interconnected system of neurons we have the model of the dynamics driving the pattern to an attractor basin and thence to convergence. But the important matter is that the process begins as soon as the partial pattern is presented and evolves under a model where the remaining part of the pattern is presented as time goes on and the dynamics allows you to reach convergence either when the full pattern is presented or when only a part is received by the system.

In addition there is a very significant result here which may be of interest. Given that the test which measures whether a computer can be treated as intelligent is the Turing test which requires a linguistic interaction between a machine and a human being in a way in which the human being cannot discern that the conversation is happening with a machine, we are effectively requiring the machine to process language at the speed at which natural language is processed (the conversation must not only be correct with respect to content and form but also with respect to latencies of response) by the 'brain'. If it turns out that the dynamics of language processing is implemented in the brain through the mechanism of quantum computations with their implied massive parallelism which allows rapid pattern processing leading to a low latency effective use of language, then a sequential digital computer, however fast it could be made under the constraints from physical laws [5] will never be able to process the sequence of instructions (algorithms) into which we have remapped natural language processing, in the right amount of time. This will lead to violations of the turing test conditions. As a consequence it may be possible to prove or find constraints to indicate that current digital computational algorithms will never capture the requisite elements of natural language processing. Which leads us to the conclusion that there will be no digital algorithmic machine which will pass the turing test. Higher order brain functions may not be simulated in a digital computer other than a quantum computer.

The questions that are important from a dynamical systems perspective are the following:

- Are these linear or non linear processes?
- Does a single neuron simultaneously participate in multiple pattern processing operations?
- Can we model these patterns as the evolution on a pattern surface?
- What does one do about the notion of time dependent dimensionality of these patterns?

9.3   The model

In order to incorporate the notion of spatiotemporal correlations existing in the universe and their subsequent evolution, as well as issues of localization of responses/processing in the brain (such as the Broca's area and so on) we need to model the brain as a physical system with the notion of spatial location and state of a neuron with the couplings between the neurons (could be in terms of quantum entanglement of states) determining the connectivity.

Although there might not be a physical scale length in the brain (this is not very obvious) and one may be able to create a scale independent model of the brain, as is usually done in digital neural modeling, it may be a sensible idea to define a physical co-ordinate systems to describe the location of the neurons, and embed this coordinate system in a coordinate system based in the physical universe. This will allow us to later discuss the issue of mapping the physical space which is a metric space into the pattern space which is most likely to be a metric space as well (as opposed to the conventional models of it being a topological space, although in neural networks a metric in terms of hamming distance or like is defined).

The metric space of neurons will have two components, the spatial separation between them and the temporal latency induced by signal travel time as usual.

If the state of a neuron (as a continuous function of space time coordinates) is defined as $N_i(r_i, \theta_i, z_i, t_i)$ the cross product between two of them can be defined as

$$C_{ij} = N_i(r_i, \theta_i, z_i, t_i)\, N_j(r_j, \theta_j, z_j, t_j). \tag{9.1}$$

The set of cross products C_{ij} between pairwise neurons in a set of N neurons is what we shall call a pattern. This pattern evolves in time in a two fold manner. The number of neurons which are part of the active pattern, N evolves as well as the cross product C_{ij}.

The driving force for this time evolution is the set of stimuli S_{ij} where i and j range from $1..P$. Thus symbolically

$$\frac{d}{dt}C_{ij} = f(S_{ij}, C_{ij})\ (i, j \Rightarrow 1..N(t)) \tag{9.2}$$

$$\frac{d}{dt}S_{ij} = g(S_{ij})\ (i, j \Rightarrow 1..P(t)). \tag{9.3}$$

In this model we assume that the base pattern can be described in terms of the set of two point correlations and higher order correlations can be described or built out of the set of two point correlations.

The pattern space evolves in the following manner. The stimulus set S_{ij} evolves in response to external information flow which is translated via the sensory system and drives the evolution of the pattern C_{ij}.

To establish the functional forms of f and g which govern the evolution we shall use the following analysis. The evolution of the stimulus set S_{ij} is expected to be linearly dependent on the external stimulus. The functional form for f which governs the evolution of the set C_{ij} is based on a fanout-fanin model, where as the stimulus set S_{ij} increases and reaches the maximum level of saturation and stays steady or decreases as a function of time.

The fan out model ensures that as the stimulus grows as a function of time, more and more neurons participate in the pattern till a maximum number of neurons is reached. This growth is combinatorical in the sense that each neuron which gets excited (or inhibited) cause further neurons to get excited in a cascade, thus quickly leading to a full pattern based evolution. Once the pattern is saturated all the neurons interact with each other in time in a self interacting model (the interactions may be either modeled classically or in terms of quantum interactions between the synapses) and potentially drive themselves to a 'steady' state (which decays rapidly). The aperiodic firing pattern of single neurons sets up resonant behavior (parameterized by the interconnection strengths between neurons). It is possible that the pattern evolves through the system reaching multiple resonances at multiple length scales. The boundary layer neurons which are instrumental in carrying the stimulus to the cluster of neurons which participated in the pattern processing would also participate and be part of the mechanism for communicating the saturation point state of the system of neurons to other parts of the brain, setting off similar behavior in them.

These localized clusters of neurons would form the subsystems which have been identified in the current brain research dealing with the notion of the brain as a weakly separated set of subsystems [6] which deal with different senses and different aspects of the activities of the brain. The localization is brought about by the fact that neurons are not symmetrically connected and are not fully connected as well.

This basic model can be used to describe most of the observed phenomena observed in terms of activities of the neural complexes. However we also need a storage or learning mechanism to complement this pattern processing to discuss how the correlations or patterns can be retained for a longer duration and can be brought to service, especially in higher order intelligent activity such as problem solving. We do not have conclusive evidence regarding problem solving and learning capabilities of lower order primates (other than human beings) but there is considerable experimental data that suggests that primate brains can store patterns to a certain degree and can make a higher order correlation extended in space and time in order to solve basic problems such as access to food.

How does one describe the storage mechanism? That is best described in the classical model of the connectivity between neurons changing as a function of pattern processing. A basic feedback loop which modify and persist the history of the patterns that have been processed by a neural complex is what serves as a memory and the ability to solve related problems. We can unify memory and the ability to recognize commonality or overlap within patterns and consequent problem solving/retrieval in terms of defining memory as an identity operation. We just recover the same pattern that was stored in the changed connectivity model which occurred when the pattern first encountered the neuron. The dynamics that processes partial data arriving in time to drive the system towards a consistent result will explain how memory can retrieve data from partial patterns.

In terms of complex problem solving, there seems to be the same feedback loop in operation, which allows problem solving techniques to be learnt by the neural system. The question is whether there is a difference between learning a technique or learning a fact in the brain. It is hypothesized that the same mechanism is in operation both in learning and in problem solving, by identifying overlap between patterns.

A neural system such as a brain is also able to use a set of neurons to simultaneously process multiple patterns and is able to superpose multiple patterns as well as process them. Different patterns may be excited by the differences in stimuli (which is a function of time) which as the stimulus grows as a function of time drives the system towards the correct response (sometimes it also causes errors in terms of confusion between nearby patterns). The details of this processing, the fan out which explores the large set of possible end patterns and the pruning effect that comes up due to dynamical processing as the stimulus grows as a function and reaches the saturation point, driving the system to the conformant state is being worked out in a following paper [7]. The mathematical model takes into account the explicit issues of the changing dimensionality of the neuronal space that constitute a particular instance of brain activity. In addition the question of whether the pattern dynamics can be modeled as a constrained evolution on a surface defined by conserved quantities are discussed therein.

The other most important aspect of this pattern processing is the ability to define a hierarchical model of processing, wherein higher order patterns can be derived from smaller patterns by a coarse averaging process in a process analogous to that of renormalization theory models. What this allows you to do is to define higher order patterns that is important in a description of the semiotic capabilities of the human brain.

9.4 Evolution of the neural system

What came first: The neuron or the brain?

One requires a fundamentally simple model in order to explain how the neural system came into being. We need to explain how the first small networks would have grown into a larger network and triggered off larger scale of pattern processing that they would have been capable of. The distinct advantage conferred by this model is the following: it is possible to encode a variety of beneficial (from the point of view of evolution) pattern processing that can be accommodated in this model easily. One does not need to deal with musical abilities or linguistic abilities or visual abilities separately in this picture. A unified approach to neural activity is afforded by this model which makes it easy to explain a systematic growth in complexity via a classical evolution as well as non linear jump in the ability corresponding to a sharp increase in brain size (number of neurons or dimensions in the patterns) as well as in terms of increased complexity of connectivity between the neurons.

The neural complexes that were initially formed would have been just capable of generating state changes in response to events happening in the external environment. Multiple external events of different kinds, say with respect to detection of light by one set of cells and detection of sound by another set of cells would have been correlated by the ganglia which would have formed at the junction of these different kinds of sensor complexes which potentially evolved into the more complex pattern processing system which we know as the brain today.

This is an important aspect as well. That of localization of patterns and how strongly/weakly patterns in one 'subsystem' can influence another. This model easily and in a natural manner explains phenomena such as music and dancing (physical movement in response to auditory signals, both voluntary and involuntary) as well as extreme correlations between unrelated subsystems in some individuals such as 'synethesia'.

It can also explain correlations between one form of sensory impulse and an apparently unrelated form of sensory response such as swirling lights triggering off an epileptic seizure. A complete description of the evolutionary track of the neural complexes and their capacity to deal with patterns of higher order of complexity is given in a separate paper.

9.5 Conclusions

In this paper, we have presented an evolutionary model for the brain as a pattern processing network of neurons in response to the observed correlations that exist in the world of space time events as gathered by the sensors embedded in the bodies of living creatures. The model describes how this large scale correlation processing system is able to use a parallel model to produce a highly efficient and low latency system that can respond to the environment as well extrapolate events in space and time, thus leading to a causal model for the Universe.

It is proposed herein that this is the only subsystem in an animal that has evolved in response to the randomness in terms of events in the universe as opposed to the deterministic behavior of the other subsystems. It is possible that the superposition (classical or quantum) that is possible in this model can effectively describe how apparently conflicting patterns and beliefs (in terms of logical paradoxes for instance [8]) can exist in the mind.

Thus this model provides a common unifying model to deal with the different aggregates of neurons that are observed in nature whose consequences still need to be explored and used to explain all the observed data.

References

[1] A. Turing, Computing Machinery and Intelligence, October, 1950.
[2] M. Sarovar, A. Ishizaki, G. R. Fleming and K. B. Whaley, Nature Physics **6**, 462 (2010).
[3] D. O. Hebb, *The Organization of Behavior* (Wiley, New York, 1949).
[4] J. Barbour, *End of Time*, (Oxford University Press, Oxford, England, 1999)
[5] S. Lloyd, Nature **406** 1047, (2008).
[6] M. Minsky, *Society of the Mind* (Simon and Schuster, New York, 1986).
[7] S. Engineer and T. Padmanabhan, *A mathematical model for neural pattern processing*, In preparation.
[8] See, http://en.wikipedia.org/wiki/Epimenides_paradox.

Articles co-authored by the contributors
with T. Padmanabhan

Compiled below are articles that have been co-authored by the contributors with Padmanabhan. Kindly note that, if an article has been co-authored by more than one contributor, it has been listed under the names of both the contributors.

Jasjeet Singh Bagla

(1) J. S. Bagla and T. Padmanabhan, *Nonlinear evolution of density perturbations using approximate constancy of gravitational potential*, MNRAS, **266**, 227 (1994).

(2) J. S. Bagla and T. Padmanabhan, *Nonlinear evolution of density perturbations*, J. Astrophys. Astron. **16**, 77 (1995).

(3) J. S. Bagla, T. Padmanabhan and J. V. Narlikar, *Crisis in cosmology: Observational constraints on Ω and H_0*, Comm. Astrophys. **18**, 275 (1996).

(4) J. S. Bagla and T. Padmanabhan, *A new statistical indicator to study nonlinear gravitational clustering and structure formation*, Ap. J. **469**, 470 (1996).

(5) J. S. Bagla and T. Padmanabhan, *Critical index and fixed point in transfer of power in nonlinear gravitational clustering*, MNRAS **286**, 1023 (1997).

(6) J. S. Bagla, B. Nath and T. Padmanabhan, *Neutral hydrogen at high redshifts as a probe of structure formation – III. Radio maps from N-body simulations*, MNRAS **289**, 671 (1997).

(7) J. S. Bagla and T. Padmanabhan, *Cosmological N-Body simulations*, Pramana – J. Phys. **42**, 161 (1997).

(8) J. S. Bagla, S. Engineer and T. Padmanabhan, *Scaling relations for gravitational clustering in two dimensions*, Ap. J. **495**, 25 (1998).

(9) J. S. Bagla, H. K. Jassal and T. Padmanabhan, *Cosmology with tachyon field as dark energy*, Phys. Rev. D **67**, 063504 (2003).

(10) H. K. Jassal, J. S. Bagla and T. Padmanabhan, *WMAP constraints on low redshift evolution of dark energy*, MNRAS **356**, L11 (2005).

(11) S. Ray, J. S. Bagla and T. Padmanabhan, *Gravitational collapse in an expanding universe: Scaling relations for two-dimensional collapse revisited*, MNRAS **360**, 546, (2005).

(12) H. K. Jassal, J. S. Bagla and T. Padmanabhan, *Observational constraints on low redshift evolution of dark energy: How consistent are different observations?*, Phys. Rev. D **72**, 103503 (2005).

(13) J. S. Bagla, G. Kulkarni and T. Padmanabhan, *Metal enrichment and reionization constraints on early star formation*, MNRAS **379**, 971 (2009).

(14) H. K. Jassal, J. S. Bagla and T. Padmanabhan, *The vanishing phantom menace*, MNRAS **405**, 2639 (2010).

Tirthankar Roy Choudhury

(1) T. Padmanabhan and T. Roy Choudhury, *The issue of choosing nothing: What determines the low energy vacuum state of nature?*, Mod. Phys. Lett. A **15**, 1813 (2000).

(2) T. Roy Choudhury, T. Padmanabhan and R. Srianand, *Semianalytic approach to understanding power spectrum of neutral hydrogen in the universe*, MNRAS **322**, 561 (2001).

(3) T. Roy Choudhury, R. Srianand and T. Padmanabhan, *Semianalytic approach to understanding the distribution of neutral hydrogen in the universe: Comparison of simulations with observations*, Ap. J. **559**, 29 (2001).

(4) T. Roy Choudhury and T. Padmanabhan, *A simple analytic model for the abundance of damped Lyman alpha absorbers*, Ap. J. **574**, 59 (2002).

(5) T. Padmanabhan and T. Roy Choudhury, *Can the clustered dark matter and smooth dark energy arise from the same scalar field?*, Phys. Rev. D **66**, 081301 (2002).

(6) T. Padmanabhan and T. Roy Choudhury, *A theoretician's analysis of the supernova data and the limitations in determining the nature of dark energy*, MNRAS **344**, 823 (2003).

(7) T. Roy Choudhury and T. Padmanabhan, *Quasi normal modes in Schwarzschild-de Sitter spacetime: A simple derivation of the level spacing of the frequencies*, Phys. Rev. D **69**, 064033 (2004).

(8) T. Roy Choudhury and T. Padmanabhan, *Cosmological parameters from supernova observations: A critical comparison of three data sets*, Astron. Astrophys. **429**, 807 (2005).

(9) T. Roy Choudhury and T. Padmanabhan, *Concept of temperature in multi-horizon spacetimes: Analysis of Schwarzschild-de Sitter*

metric, Gen. Rel. Grav. **39**, 1789 (2007).

Nissim Kanekar

(1) T. Padmanabhan and N. Kanekar, *Gravitational clustering in a D-dimensional universe*, Phys. Rev. D **61**, 023515 (2000).
(2) S. Engineer, N. Kanekar and T. Padmanabhan, *Nonlinear density evolution from an improved spherical collapse model*, MNRAS **314**, 279 (2000).
(3) N. Kanekar and T. Padmanabhan, *The effects of anti-correlation on gravitational clustering*, MNRAS **324**, 988 (2001).

Aseem Paranjape

(1) A. Paranjape, S. Sarkar and T. Padmanabhan, *Thermodynamic route to field equations in Lanczos-Lovelock Gravity*, Phys. Rev. D **74**, 104015 (2006).
(2) T. Padmanabhan and A. Paranjape, *Entropy of null surfaces and dynamics of spacetime*, Phys. Rev. D **75**, 064004 (2007).
(3) A. Paranjape and T. Padmanabhan, *Radiation from collapsing shells, semiclassical backreaction and black hole formation*, Phys. Rev. D **80**, 044011 (2009).

T. R. Seshadri

(1) T. Padmanabhan and T. R. Seshadri, *Probing the origin of large inhomogeneities in inflation using a toy quantum mechanical model*, Phys. Rev. D **34**, 951 (1986).
(2) T. Padmanabhan, T. R. Seshadri and T. P. Singh, *Uncertainty principle and the quantum fluctuations of the Schwarzschild light cones*, Int. J. Mod. Phys. A **1**, 491 (1986).
(3) T. Padmanabhan and T. R. Seshadri, *Horizon problem and inflation*, J. Astrophys. Astron. **8**, 275 (1987).
(4) T. Padmanabhan and T. R. Seshadri, *Uncertainty principle and the horizon size of our universe*, Gen. Rel. Grav. **19**, 791 (1987).
(5) T. Padmanabhan and T. R. Seshadri, *Does inflation solve the horizon problem?*, Class. Quant. Grav. **5**, 221 (1988).

(6) T. Padmanabhan and T. R. Seshadri, *Quantum uncertainty in the horizon size in an inflationary universe*, Int. J. Mod. Phys. A **3**, 2113 (1988).

(7) T. Padmanabhan, T. R. Seshadri and T. P. Singh, *Making inflation work: Damping of density perturbations due to Planck energy cutoff*, Phys. Rev. D **39**, 2100 (1989).

(8) T. R. Seshadri and T. Padmanabhan, *Gaussian states in the de-Sitter spacetime and the evolution of semiclassical density perturbations: 1. Homogeneous mode*, J. Astrophys. Astron. **10**, 391 (1989).

(9) T. R. Seshadri and T. Padmanabhan, *Gaussian states in the de-Sitter spacetime and the evolution of semiclassical density perturbations: 2. Inhomogeneous modes*, J. Astrophys. Astron. **10**, 407 (1989).

S. Shankaranarayanan

(1) S. Shankaranarayanan and T. Padmanabhan, *Hypothesis of path integral duality: Applications to QED*, Int. J. Mod. Phys. **10**, 351 (2001).

(2) S. Shankaranarayanan, K. Srinivasan and T. Padmanabhan, *Method of complex paths and general covariance of Hawking radiation*, Mod. Phys. Letts. **16**, 571 (2001).

(3) T. Padmanabhan and S. Shankaranarayanan, *Vanishing of cosmological constant in nonfactorizable geometry*, Phys. Rev. D. **63**, 105021 (2001).

(4) S. Shankaranarayanan, T. Padmanabhan and K. Srinivasan, *Hawking radiation in different coordinate settings: Complex paths approach*, Class. Quant. Grav. **19**, 2671 (2002).

(5) D. Kothawala, L. Sriramkumar, S. Shankaranarayanan and T. Padmanabhan, *Path integral duality modified propagators in spacetimes with constant curvature*, Phys. Rev. D **79**, 104020 (2009).

T. P. Singh

(1) T. Padmanabhan, T. R. Seshadri and T. P. Singh, *Uncertainty principle and the quantum fluctuations of Schwarzschild light cones*, Int. J. Mod. Phys. A **1**, 491 (1986).

(2) T. P. Singh and T. Padmanabhan, *Semiclassical cosmology with a scalar field*, Phys. Rev. D **35**, 2993 (1987).

(3) T. Padmanabhan and T. P. Singh, *Response of an accelerated detector coupled to the stress-energy tensor*, Class. Quant. Grav. **4**, 1397 (1987).

(4) T. P. Singh and T. Padmanabhan, *An attempt to explain the smallness of the cosmological constant*, Int. J. Mod. Phys. A **3**, 1593 (1988).

(5) T. Padmanabhan and T. P. Singh, *Response of accelerated detectors in coherent states and the semi-classical limit*, Phys. Rev. D **38**, 2457 (1988).

(6) A. R. Janah, T. Padmanabhan and T. P. Singh, *On Feynman's formula for the electromagnetic field of an arbitrarily moving charge*, Am. J. Phys. **1036**, 56 (1988).

(7) T. Padmanabhan, T. R. Seshadri and T. P. Singh, *Making inflation work: Damping of density perturbations due to Planck energy cutoff*, Phys. Rev. D **39**, 2100 (1989).

(8) T. P. Singh and T. Padmanabhan, *Notes on semiclassical gravity*, Ann. Phys. **196**, 296 (1989).

(9) T. Padmanabhan and T. P. Singh, *On the semiclassical limit of Wheeler-DeWitt equation*, Class. Quant. Grav. **7**, 441 (1990).

(10) C. Keifer, T. Padmanabhan and T. P. Singh, *A comparison between semiclassical gravity and semiclassical electrodynamics*, Class. Quant. Grav. **8**, L185 (1991).

(11) T. Padmanabhan and T. P. Singh, *A comparison of various approaches to the back reaction problem*, Ann. Phys. **221**, 217 (1992).

(12) T. Padmanabhan and T. P. Singh, *A note on the thermodynamics of gravitational radiation*, Class. Quant. Grav. **20**, 4419 (2003).

L. Sriramkumar

(1) L. Sriramkumar and T. Padmanabhan, *Finite time response of inertial and uniformly accelerated Unruh-DeWitt detectors*, Class. Quant. Grav. **13**, 2061 (1996).

(2) L. Sriramkumar and T. Padmanabhan, *Does a non-zero tunnelling probability imply particle production in time independent electromagnetic backgrounds*, Phys. Rev. D **54**, 7599 (1996).

(3) L. Sriramkumar, R. Mukund and T. Padmanabhan, *Non-trivial*

classical backgrounds with vanishing quantum corrections, Phys. Rev. D **55**, 6147 (1997).

(4) K. Srinivasan, L. Sriramkumar and T. Padmanabhan, *Plane waves viewed from an accelerated frame: Quantum physics in classical setting*, Phys. Rev. D **56**, 6692 (1997).

(5) K. Srinivasan, L. Sriramkumar and T. Padmanabhan, *Possible quantum interpretation of certain power spectra in classical field theory*, Int. J. Mod. Phys. D **6**, 607 (1997).

(6) K. Srinivasan, L. Sriramkumar and T. Padmanabhan, *The hypothesis of path integral duality II: Corrections to quantum field theoretic results*, Phys. Rev. D. **58**, 044009 (1998).

(7) L. Sriramkumar and T. Padmanabhan, *Probes of vacuum structure of quantum fields in classical backgrounds*, Int. J. Mod. Phys. D **11**, 1, (2002).

(8) L. Sriramkumar and T. Padmanabhan, *Initial state of matter fields and trans-Planckian physics: Can CMB observations disentangle the two?*, Phys. Rev. D **71**, 103512 (2005).

(9) D. Kothawala, L. Sriramkumar, S. Shankaranarayanan and T. Padmanabhan, *Path integral duality modified propagators in spacetimes with constant curvature*, Phys. Rev. D **79**, 104020 (2009).

Sunu Engineer

(1) T. Padmanabhan and S. Engineer, *Nonlinear gravitational clustering: Dreams of a paradigm*, Ap. J. **493**, 509 (1998).

(2) J. S. Bagla, S. Engineer and T. Padmanabhan, *Scaling relations for gravitational clustering in two dimensions*, Ap. J. **495**, 25 (1998).

(3) S. Engineer, K. Srinivasan and T. Padmanabhan, *Formal analysis of two dimensional gravity*, Ap. J. **512**, 1 (1999).

(4) S. Engineer, N. Kanekar and T. Padmanabhan, *Nonlinear density evolution from an improved spherical collapse model*, MNRAS **314**, 279 (2000).

Index